PERGAMON INTERNATIONAL LIBRARY
of Science, Technology, Engineering and Social Studies

*The 1000-volume original paperback library in aid of education,
industrial training and the enjoyment of leisure*
Publisher: Robert Maxwell, M. C.

Principles of
Water Quality Control

THIRD EDITION

(revised and enlarged)

THE PERGAMON TEXTBOOK
INSPECTION COPY SERVICE

An inspection copy of any book published in the Pergamon International Library will gladly be sent
to academic staff without obligation for their consideration for course adoption or recommendation.
Copies may be retained for a period of 60 days from receipt and returned if not suitable. When a
particular title is adopted or recommended for adoption for class use and the recommendation results
in a sale of 12 or more copies, the inspection copy may be retained with our compliments. The
Publishers will be pleased to receive suggestions for revised editions and new titles to be published in
this important International Library.

Other Pergamon Titles of Interest

Pergamon Related Journals (*Free specimen copy gladly sent on request*)

Principles of
Water Quality Control

THIRD EDITION

by

T. H. Y. TEBBUTT

B.Sc., S.M., Ph.D., M.I.C.E., M.I.W.E.S., F.I.P.H.E.,
F.Inst.W.P.C., M.A.S.C.E.

Department of Civil Engineering, University of Birmingham

PERGAMON PRESS

OXFORD · NEW YORK · TORONTO · SYDNEY · PARIS · FRANKFURT

U.K.	Pergamon Press Ltd., Headington Hill Hall, Oxford OX3 0BW, England
U.S.A.	Pergamon Press Inc., Maxwell House, Fairview Park, Elmsford, New York 10523, U.S.A.
CANADA	Pergamon Press Canada Ltd., Suite 104, 150 Consumers Rd., Willowdale, Ontario M2J 1P9, Canada
AUSTRALIA	Pergamon Press (Aust.) Pty. Ltd., P.O. Box 544, Potts Point, N.S.W. 2011, Australia
FRANCE	Pergamon Press SARL, 24 rue des Ecoles, 75240 Paris, Cedex 05, France
FEDERAL REPUBLIC OF GERMANY	Pergamon Press GmbH, Hammerweg 6, D-6242 Kronberg-Taunus, Federal Republic of Germany

First Edition 1971, Reprinted 1975
Second Edition 1977, Reprinted 1979
Third Edition 1983

Library of Congress Cataloging in Publication Data

Tebbutt, T. H. Y.
Principles of water quality control.
(Pergamon international library of science, technology, engineering, and social studies)
Includes index.
1. Water quality management.   2. Water—
Purification.
I. Title.   II. Series.
TD365.T4   1982   628.1′6   82-9863

British Library Cataloguing in Publication Data

Tebbutt, T. H. Y.
Principles of water quality control.—3rd ed.
1. Water quality management
I. Title
628.1′61   TD365

ISBN 0-08-028705-0 Hard cover
ISBN 0-08-028704-2 Flexi cover

Printed in Great Britain by A. Wheaton & Co. Ltd., Exeter

Preface to the Third Edition

IN THE period since the first publication of this book public awareness of environmental pollution problems has grown although of late the world economic situation has tended to limit expenditure on environmental protection. To some extent such financial constraints are not altogether undesirable since in their absence costly control measures with little real benefit may be implemented. In any decision-making process it is important that the basic facts are fully understood. In the hope of aiding the proper understanding of water quality control measures the text has been fairly extensively modified to include consideration of current philosophies and techniques. The vital role of water in the developing countries has been recognized by the UN International Drinking Water Supply and Sanitation Decade and two new chapters, on Water Quality and Health and Water Supply and Sanitation in Developing Countries, have been added to the text. These cannot cover all the particular problems of developing countries but aim to provide an introduction to the topic. The recommendations for further reading have been expanded to encourage the reader to develop understanding of the basic concepts and gain a wider appreciation of the topics covered.

As before, the book is primarily intended as a text for undergraduate students in civil engineering and as preliminary reading for graduate students in relevant study areas. It should also be useful for readers wishing to obtain an understanding of water quality control for post-experience training and professional development.

Department of Civil Engineering,   T.H.Y.T.
University of Birmingham
1982

Preface to the First Edition

THIS book is designed as a text for undergraduate civil engineering courses and as preliminary reading for postgraduate courses in public health engineering and water resources technology. It is hoped that it will also be of value to workers already in the field and to students preparing for the examinations of the Institute of Water Pollution Control and the Institution of Public Health Engineers. The text is based on my own lecture courses to undergraduate civil engineers augmented by material prepared for extramural short courses. Wherever possible, simple illustrations have been used to clarify the text. Reproduction of detailed working drawings has been deliberately avoided since in my experience these are often confusing to the student until the fundamentals of the subject are fully understood. Problems with answers have been included throughout the book so that the reader can check his understanding of the text. The SI system of units has been adopted and it is hoped that the complete absence of Imperial units will encourage familiarity with their metric replacements.

The preparation of this book owes much to the enthusiasm for the subject which I gained from my first mentor, Professor P. C. G. Isaac of the University of Newcastle upon Tyne. I am most grateful to my colleague M. J. Hamlin for his helpful comments on the text and to Vanessa Green for her expert typing of the manuscript.

Department of Civil Engineering   T.H.Y.T.
University of Birmingham

Contents

CHAPTER 1

Introduction

WATER is probably the most important natural resource in the world since without it life cannot exist and industry cannot operate. Unlike many other raw materials there is no substitute for water in many of its uses. Water plays a vital role in the development of communities since a reliable supply of water is an essential prerequisite for the establishment of a permanent community. Unfortunately, the liquid and solid wastes from such a community have a considerable potential for environmental pollution. In primitive civilizations the remedy for this pollution problem was simply to move the community to another suitable site. In more advanced civilizations such upheavals become impracticable and measures must be taken to protect and augment water supplies and for the satisfactory disposal of waste materials. The concept of water as a natural resource which must be carefully managed is very necessary as growing populations and industrial developments demand ever-increasing supplies of water.

The importance of water supply and sanitation provisions was recognized centuries ago in ancient civilizations. Archaeological evidence shows the existence of latrines and drains in Neolithic dwellings and the Minoan civilization in Crete 2000 years B.C. had clay water and sewage pipes with flushing toilets in the houses. The Romans had highly developed water supply and drainage systems and their cities used large amounts of water with continuously operating fountains being a major source of supply for the majority of the population although wealthy families had their own piped supplies. Large aqueducts, some of which still remain, were constructed over distances up to 80 km to bring adequate supplies of good-quality water into the cities. Stone sewers in the streets removed surface-water and collected the discharges from latrines for conveyance beyond the city limits. With the demise of the Roman Empire most of their public works installations fell into disuse and for centuries water supply and sanitation provisions were virtually non-existent. In the Middle Ages, towns started to develop at important crossing points on rivers and these rivers usually provided a convenient source of water and an apparently convenient means of waste disposal. Although sewers were built in the larger towns they were intended solely for the removal of surfacewater and in the UK the discharge of foul sewage to the sewers was forbidden by law until 1815. Sanitary provisions were usually minimal: in 1579 one street in London with sixty houses had three communal latrines. Discharges of liquid and solid wastes from windows into the street were common and it is not surprising that life expectancy was less than half the current figure in the developed world. In an attempt to improve matters a law was passed in 1847 which made it obligatory in London for cesspit and latrine wastes to be discharged to the sewers. London's sewers drained to the Thames, from which much of the city's water was obtained, and in addition the poor state of repair of many of the sewers allowed the contents to leak into the aquifer which was the other main source of water. The inevitable consequences of this state of affairs were that water sources became increasingly contaminated by sewage, the Thames became objectionable to both sight and smell, and most seriously, waterborne diseases became rampant in the city. The Broad Street Pump outbreak of cholera in 1854 which caused 10 000 deaths provided the evidence for Dr. John Snow to demonstrate the connection between sewage pollution of water and enteric diseases like cholera and typhoid. Public outcry resulted in the commissioning of the first major public health engineering works of modern times; Bazalgette's intercepting sewers which collected sewage discharges and conveyed them downstream of London for discharge to the estuary, and water abstraction from Teddington on the non-tidal part of the river. Thus by 1870 waterborne outbreaks had been largely brought under control in the UK and similar developments were taking place in Western Europe and the cities of the USA. The Industrial Revolution greatly increased the urban water demand and the late nineteenth century saw the construction of major water-supply schemes involving large upland impoundments of which the Elan Valley scheme for Birmingham and the Croton and Catskill reservoirs serving New York are examples.

Only by continual and costly attention to water quality control has it been possible to virtually eradicate waterborne diseases from developed countries. Such achievements must not, however, be allowed to mask the appalling situation regarding water supply and sanitation in much of the developing world. A survey in 1975 found that 80% of the world's rural population and 23% of the urban population had no reasonable access to a safe water supply. The sanitation situation was even worse with 85% of the rural population and 25% of the urban population having no sanitary provision at all. The growth of population in developing countries, due to the high birth rate, is such that unless strenuous efforts to increase water supply and sanitation facilities are made, the percentage of the world's population with satisfactory facilities would actually decrease in the future. The United Nations Organization has therefore designated the period 1981–90 as the International Drinking Water Supply and Sanitation Decade with the aim of providing safe water and adequate sanitation for all. Such a target will be difficult to achieve, involving as it must large investments in finance and manpower, although it has been estimated that the cost of meeting the Decade target would be about 5% of global spending on armaments during the same period.

In developed countries, demands for water are now fairly static and basic water quality-control measures are well established. However, many of the existing water-supply and sewage schemes are now relatively old so that their reconstruction will pose problems in the future. In addition, as knowledge of the effects of all forms of environmental pollution increases so new potential hazards appear, for example there is current concern about the possible carcinogenic hazards arising from the presence of minute concentrations of some organic compounds in water. Thus, throughout the world, various aspects of water quality control will continue to be vital in safeguarding public health.

1.1. The Engineer's Role

Civil engineers have traditionally been concerned with water supply and wastewater disposal and indeed public health engineering probably forms the largest single activity of the civil engineering profession. Water science and technology is, however, an interdisciplinary topic involving the application of biological, chemical and physical principles in association with engineering techniques. The public health engineer in co-operation with scientific colleagues plays a major role in reducing the incidence of many water-related diseases.

The responsibilities of the engineer start with the development of water sources to provide an ample supply of water of wholesome quality, i.e. a water free from

> visible suspended matter,
> excessive colour,
> taste and odour,
> objectionable dissolved matter,
> aggressive constituents,
> bacteria indicative of faecal pollution.

For drinking water supplies the water must obviously be fit for human consumption, i.e. potable, and it should also be palatable, i.e. aesthetically attractive. In addition, as far as is feasible, it should be suitable for other domestic uses such as clothes washing, etc.

There is a vast amount of water present in the earth and its surrounding atmosphere and about 7% of the earth's mass is made up of water. However, 97% of all the water is found as saline water in the oceans and much of the remaining 3% which exists as freshwater is incorporated in the polar ice caps. Thus only about 0.7% of the earth's water occurs in freshwater lakes and rivers, in accessible aquifers and in the atmosphere. Even so, if the water were evenly distributed on the earth's surface and if the human population were similarly distributed there would be no shortages of water. Because the spatial distribution of rainfall is uneven and densely populated urban areas consume large amounts of water, shortages do occur. The science of hydrology which is concerned with the management of the hydrological cycle and its water resources plays an important role in satisfying ever-increasing demands for water throughout the developing world. It must therefore be appreciated that in considerations of water resources it is essential to assess both the quality and quantity of the source.

Having provided a suitable quality and quantity of water by development of a source and construction of treatment processes, the treated water must be delivered to the consumers via a complex distribution system. Domestic and industrial uses of water generally produce a deterioration in quality and the wastewaters must be collected and given effective treatment before release to the environment. In many cases treated wastewaters become a not insignificant part of the water resource and are available for reuse. The basic elements of a water resource system are shown in Fig. 1.1.

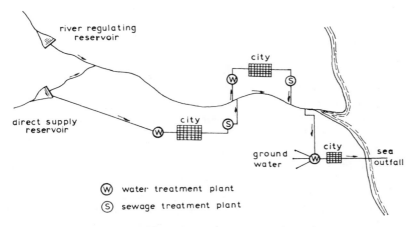

FIG. 1.1. Water supply and wastewater disposal.

Further Reading

BINNIE, G. M., *Early Victorian Water Engineers*, Thomas Telford Ltd., London, 1981.

HARTLEY, D., *Water in England*, Macdonald & Co. (Publishers) Ltd., London, 1964.

HARTLEY, SIR HAROLD, The engineer's contribution to the conservation of natural resources. *Proc. Instn Civ. Engrs*, **4**, 1955, 692.

ISAAC, P. C. G., Roman public health engineering. *Proc. Instn Civ. Engrs*, **68**, 1980, part 1, 215.

SMITH, K., *Water in Britain*, Macmillan & Co. Ltd., London, 1972.

WRIGHT, L., *Clean and Decent*, Routledge & Kegan Paul Ltd., London, 1960.

CHAPTER 2

Characteristics of Waters and Wastewaters

ALTHOUGH water is normally considered as H_2O, all natural waters contain varying amounts of other materials in concentrations ranging from a few milligrams per litre in rain to about 35 000 mg/l in seawater. Wastewaters usually contain most of the constituents of the water supply to the area with additional impurities from the waste producing process. Thus, man produces about 6 g of chloride per day so that with a water consumption of 200 l/person/day the sewage contains 30 mg/l more chloride than the domestic water. An average raw sewage contains around 1000 mg/l of solids in solution and suspension and is thus about 99.9 % pure water. Clearly a simple measure of the total solid content of a sample is insufficient to specify its character since a clear sparkling ground water might have the same total solids content as a raw sewage. To gain a true picture of the nature of a particular sample it is often necessary to measure several different properties by carrying out an analysis under the broad headings of physical, chemical and biological characteristics.

Not all characteristics would be investigated for any one sample and Table 2.1 shows the properties most likely to be measured for various samples.

2.1. Physical Characteristics

Physical properties are in many cases relatively easy to measure and some may be readily observable by a layman.

1. *Temperature.* Basically important for its effect on other properties, e.g. speeding up of chemical reactions, reduction in solubility of gases, amplification of tastes and odours, etc.

2. *Taste and odour.* Due to dissolved impurities, often organic in nature,

TABLE 2.1. IMPORTANT CHARACTERISTICS FOR VARIOUS SAMPLES

Characteristic	River water	Drinking water	Raw sewage	Sewage effluent
pH	×	×	×	×
Temperature	×	×	×	
Colour	×	×		
Turbidity	×	×		
Taste		×		
Odour	×	×		
Total solids	×	×		
Settleable solids			×	
Suspended solids			×	×
Conductivity	×	×		
Radioactivity	×	×		
Alkalinity	×	×	×	×
Acidity	×	×	×	×
Hardness	×	×		
DO	×	×		
BOD	×		×	×
PV, COD or TOC	×		×	×
Organic nitrogen			×	×
Ammonia nitrogen	×		×	×
Nitrite nitrogen	×		×	×
Nitrate nitrogen	×	×	×	×
Chloride	×	×		
Phosphate	×		×	×
Synthetic detergent	×		×	×
Bacteriological counts	×	×		

e.g. phenols and chlorophenols. They are subjective properties which are difficult to measure.

3. *Colour.* Even pure water is not colourless; it has a pale green-blue tint in large volumes. It is necessary to differentiate between true colour due to material in solution and apparent colour due to suspended matter. Natural yellow colour in water from upland catchments is due to organic acids which are not in any way harmful, being similar to tannic acid from tea. Nevertheless, many consumers object to a highly coloured water on aesthetic grounds and coloured waters may be unacceptable for certain industrial uses, e.g. production of high-grade art papers.

4. *Turbidity.* The presence of colloidal solids gives liquid a cloudy appearance which is aesthetically unattractive and may be harmful.

Turbidity in water may be due to clay and silt particles, discharges of sewage or industrial wastes, or to the presence of large numbers of micro-organisms.

5. *Solids.* These may be present in suspension and/or in solution and they may be divided into organic matter and inorganic matter. Total dissolved solids (TDS) are due to soluble materials whereas suspended solids (SS) are discrete particles which can be measured by filtering a sample through a fine paper. Settleable solids are those removed in a standard settling procedure using a 1-l cylinder. They are determined from the difference between SS in the supernatant and the original SS in the sample.

6. *Electrical conductivity.* The conductivity of a solution depends on the quantity of dissolved salts present and for dilute solutions it is approximately proportional to the TDS content:

$$K = \frac{\text{conductivity (S/m)}}{\text{TDS (mg/l)}} \tag{2.1}$$

Knowing the appropriate value of K for a particular water, the measurement of conductivity provides a rapid indication of TDS content.

2.2. Chemical Characteristics

Chemical characteristics tend to be more specific in nature than some of the physical parameters and are thus more immediately useful in assessing the properties of a sample.

It is useful at this point to set out some basic chemical definitions:

Atomic weight—weight of an atom of an element referred to a standard based on the carbon isotope C^{12}.

Molecular weight—total atomic weight of all atoms in a molecule.

Molar solution—solution containing the g molecular weight of the substance in 1 litre.

Valence—property of an element measured by the number of atoms of hydrogen that one atom of the element can hold in combination or displace

$$\text{Equivalent weight} = \frac{\text{Molecular weight}}{z} \tag{2.2}$$

where z = for acids, the number of moles of H^+ obtainable from 1 mole of acid.

    = for bases, the number of moles of H^+ with which 1 mole of base will react.

(A mole is the molecular weight in g.)

Normal (N) *solution*—solution containing the g equivalent weight of the substance in 1 litre.

Some important chemical characteristics are described below.

1. *pH*. The *intensity* of acidity or alkalinity of a sample is measured on the pH scale which actually measures the concentration of hydrogen ions present.

Water is only weakly ionized:

$$H_2O \rightleftharpoons H^+ + OH^-$$

Since only about 10^{-7} molar concentrations of $[H^+]$ and $[OH^-]$ are present at equilibrium $[H_2O]$ may be taken as unity. Thus

$$[H^+][OH^-] = K = 1.01 \times 10^{-14} \text{ mole/l at } 25°C \qquad (2.3)$$

Since this relationship must be satisfied for all dilute aqueous solutions the acidic or basic nature of the solution can be specified by one parameter—the concentration of hydrogen ions. This is conveniently expressed by the function pH

$$pH = -\log_{10}[H^+] = \log_{10}\frac{1}{[H^+]} \qquad (2.4)$$

resulting in a scale from 0 to 14 with 7 as neutrality, below 7 being acid and above 7 being alkaline.

Many chemical reactions are controlled by pH and biological activity is usually restricted to a fairly narrow pH range of 6–8. Highly acidic or highly alkaline waters are undesirable because of corrosion hazards and possible difficulties in treatment.

2. *Oxidation-reduction potential (ORP)*. In any system undergoing oxidation there is a continual change in the ratio between the materials in the reduced form and those in the oxidized form. In such a situation the potential required to transfer electrons from the oxidant to the reductant is approximated by:

$$ORP = E° - \frac{0.059}{z}\log_{10}\frac{[\text{products}]}{[\text{reactants}]} \qquad (2.5)$$

where E° = cell oxidation potential referred to H = 0,

z = number of electrons in the reaction.

Operational experience has established ORP values likely to be critical for various oxidation reactions. Aerobic reactions show ORP values of $> +200\,mV$, anaerobic reactions occur below $+50\,mV$.

3. *Alkalinity*. Due to the presence of bicarbonate HCO_3^-, carbonate $CO_3^=$, or hydroxide OH^-. Most of the natural alkalinity in waters is due to HCO_3^- produced by the action of ground water on limestone or chalk:

$$CaCO_3 + H_2O + CO_2 \rightarrow Ca(HCO_3)_2$$
insoluble   from   soluble
soil
bacteria

Alkalinity is useful in waters and wastes in that it provides buffering to resist changes in pH. It is normally divided into caustic alkalinity above pH 8.2 and total alkalinity above pH 4.5. Alkalinity can exist down to pH 4.5 because of the fact that HCO_3^- is not completely neutralized until this pH is reached. The amount of alkalinity present is expressed in terms of $CaCO_3$. (See also p. 22.)

4. *Acidity*. Most natural waters and domestic sewage are buffered by a CO_2–HCO_3^- system. Carbonic acid H_2CO_3 is not fully neutralized until pH 8.2 and will not depress the pH below 4.5. Thus CO_2 acidity is in the pH range 8.2 to 4.5, mineral acidity (almost always due to industrial wastes) occurs below pH 4.5. Acidity is expressed in terms of $CaCO_3$.

5. *Hardness*. This is the property of a water which prevents lather formation with soap and produces scale in hot-water systems. It is due mainly to the metallic ions Ca^{++} and Mg^{++} although Fe^{++} and Sr^{++} are also responsible. The metals are usually associated with HCO_3^-, $SO_4^=$, Cl^-, and NO_3^-. There is no health hazard, but economic disadvantages of a hard water include increased soap consumption and higher fuel costs. Hardness is expressed in terms of $CaCO_3$ and is divided into two forms:

(a) Carbonate hardness—metals associated with HCO_3^-.

(b) Non-carbonate hardness—metals associated with $SO_4^=$, Cl^-, NO_3^-.

Total hardness—alkalinity = non-carbonate hardness.

If high concentrations of Na and K salts are present, the non-carbonate hardness may be negative since such salts could form alkalinity without producing hardness.

6. *Dissolved Oxygen (DO)*. Oxygen is a most important element in water quality control. Its presence is essential to maintain the higher forms of biological life and the effect of a waste discharge on a river is largely determined by the oxygen balance of the system. Unfortunately oxygen is only slightly soluble in water.

Temp, $°C$	0	10	20	30
DO, mg/l	14.6	11.3	9.1	7.6

Clean surface waters are normally saturated with DO, but such DO can be rapidly removed by the oxygen demand of organic wastes. Game fish require at least 5 mg/l DO and coarse fish will not exist below about 2 mg/l DO. Oxygen-saturated waters have a pleasant taste and waters lacking in DO have an insipid taste; drinking waters are thus aerated if necessary to ensure maximum DO. For boiler feed waters DO is undesirable because its presence increases the risk of corrosion.

7. *Oxygen demand*. Organic compounds are generally unstable and may be oxidized biologically or chemically to stable, relatively inert, end products such as CO_2, NO_3, H_2O. An indication of the organic content of a waste can be obtained by measuring the amount of oxygen required for its stabilization:

    (a) *Biochemical oxygen demand (BOD)*—is a measure of the oxygen required by micro-organisms whilst breaking down organic matter.

    (b) *Permanganate value (PV)*—chemical oxidation using potassium permanganate solution.

    (c) *Chemical oxygen demand (COD)*—chemical oxidation using boiling potassium dichromate and concentrated sulphuric acid.

The magnitude of the results obtained is usually PV < BOD < COD.

Organic matter may be determined directly as total organic carbon (*TOC*) by specialized combustion techniques or by using the UV absorption characteristics of the sample. In both cases commercial instruments are readily available but are relatively costly to buy and operate.

Because of the importance of oxygen demand considerations, the subject is covered in detail in Chapter 6.

8. *Nitrogen*. This is an important element since biological reactions can

only proceed in the presence of sufficient nitrogen. Nitrogen exists in four main forms as far as public health engineering is concerned:

(a) Organic nitrogen—nitrogen in the form of proteins, amino acids and urea.

(b) Ammonia nitrogen—nitrogen as ammonium salts, e.g. $(NH_4)_2 CO_3$, or as free ammonia.

(c) Nitrite nitrogen—an intermediate oxidation stage not normally present in large amounts.

(d) Nitrate nitrogen—final oxidation product of nitrogen.

Oxidation of nitrogen compounds, termed nitrification, proceeds thus:

$$Org.N + O_2 \rightarrow Amm.N + O_2 \rightarrow NO_2 - N + O_2 \rightarrow NO_3 - N$$

Reduction of nitrogen, termed denitrification, may reverse the process:

$$NO_3^- \rightarrow NO_2^- \overset{\nearrow NH_3}{\underset{\searrow N_2}{}}$$

The relative concentrations of the different forms of nitrogen give a useful indication of the nature and strength of the sample. Before the availability of bacteriological analysis the quality of waters was often assessed by considering the nitrogen content. A water containing high Org.N and Amm.N with little $NO_2 - N$ and $NO_3 - N$ would be considered unsafe because of recent pollution. On the other hand, a sample with no Org.N and Amm.N and some $NO_3 - N$ would be considered safe as nitrification had occurred and thus pollution could not have been recent.

9. *Chloride.* Responsible for brackish taste in water and is an indicator of sewage pollution because of the chloride content of urine. Threshold level for Cl^- taste is 250–500 mg/l, although up to 1500 mg/l is unlikely to be harmful to healthy consumers.

Many other specialized chemical characteristics may be assessed when dealing with industrial wastewaters, e.g. toxic metals, cyanide, phenol, oils and greases, etc.

2.3. Biological Characteristics

The subject of microbiology will be discussed in Chapter 4 and it suffices here to say that bacteriological analysis of water supplies usually provides the most sensitive quality parameter.

Almost all organic wastes contain large numbers of micro-organisms, sewage containing over 10^6/ml, but the actual numbers present are not often determined. After conventional sewage treatment the effluent still contains large numbers of micro-organisms as do many natural surface waters.

2.4. Typical Characteristics

Since waters and wastewaters vary widely in their characteristics it is not really possible to give details of what might be termed normal characteristics for a particular sample. As a guide, however, Table 2.2 gives typical analyses for various types of water source and Table 2.3 gives analyses of domestic sewage before and after treatment. A diagrammatic representation of the nature of sewage appears in Fig. 2.1.

In the case of water used for potable supply it is common practice to assess its quality in relation to specified guidelines or standards. The formulation of such guide values requires critical assessment of the

TABLE 2.2. CHARACTERISTICS OF VARIOUS WATER SOURCES

Characteristic mg/l (except where noted)	Source		
	Upland catchment	Lowland river	Chalk aquifer
pH (units)	6.0	7.5	7.2
Total solids	50	400	300
Conductivity (μS/cm)	45	700	600
Chloride	10	50	25
Alkalinity (total)	20	175	110
Hardness (total)	10	200	200
Colour (°H)	70	40	< 5
Turbidity (NTU)	5	50	< 5
Amm.N	0.05	0.5	0.05
NO_3N	0.1	2.0	0.5
DO (percent saturation)	100	75	2
BOD	2	4	2
22°C Colonies/ml	100	30 000	10
37°C Colonies/ml	10	5000	5
Coliform MPN/100 ml	20	20 000	5

TABLE 2.3. TYPICAL SEWAGE ANALYSES

Characteristic mg/l	Source		
	Crude	Settled	Final effluent
BOD	300	175	20
COD	700	400	90
TOC	200	90	30
SS	400	200	30
Amm.N	40	40	5
NO_3N	< 1	< 1	20

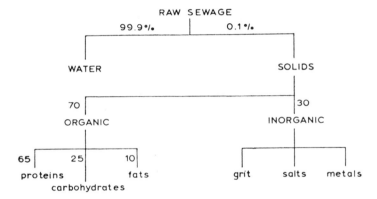

FIG. 2.1. Composition of sewage.

properties of the various constituents and it is often useful to classify constituents into five groups.

1. Organoleptic parameters—these are characteristics readily observable by the consumer but usually having little health significance; typical examples are colour, turbidity, taste and odour.

2. Natural physico-chemical parameters—these are normal characteristics of waters such as pH, conductivity, total solids, alkalinity, hardness, dissolved oxygen, etc. A few of these parameters may have health significance but in general the aim in setting guidelines is to prevent the supply of excessively unbalanced waters.

TABLE 2.4. EEC STANDARDS FOR SURFACEWATERS USED FOR POTABLE ABSTRACTIONS*

Treatment type	A1		A2		A3	
Parameter (mg/l except where noted)	Guide limit	Mandatory limit	Guide limit	Mandatory limit	Guide limit	Mandatory limit
pH units	6.5–8.5		5.5–9.0		5.5–9.0	
Colour units	10	20	50	100	50	200
Suspended solids	25					
Temperature, °C	22	25	22	25	22	25
Conductivity (μS/cm)	1000		1000		1000	
Odour TON	3		10		20	
Nitrate (as NO_3)	25	50		50		50
Fluoride	0.7–1.0	1.5	0.7–1.7		0.7–1.7	
Iron (soluble)	0.1	0.3	1.0	2.0	1.0	
Manganese	0.05		0.1		1.0	
Copper	0.02	0.05	0.05		1.0	
Zinc	0.5	3.0	1.0	5.0	1.0	5.0
Boron	1.0		1.0		1.0	
Arsenic	0.01	0.05		0.05	0.05	0.1
Cadmium	0.001	0.005	0.001	0.005	0.001	0.005
Chromium (total)		0.05		0.05		0.05
Lead		0.05		0.05		0.05
Selenium		0.04		0.01		0.01
Mercury	0.0005	0.001	0.0005	0.001	0.0005	0.001
Barium		0.1		1.0		1.0
Cyanide		0.05		0.05		0.05
Sulphate	150	250	150	250	150	250
Chloride	200		200		200	
MBAS	0.2		0.2		0.5	
Phosphate (as P_2O_5)	0.4		0.7		0.7	
Phenol		0.001	0.001	0.005	0.01	0.1
Hydrocarbons (ether soluble)		0.05		0.2	0.5	1.0

TABLE 2.4 (contd.)

Treatment type	A1		A2		A3	
Parameter (mg/l except where noted)	Guide limit	Mandatory limit	Guide limit	Mandatory limit	Guide limit	Mandatory limit
PAH		0.0002		0.0002		0.001
Pesticides		0.001		0.0025		0.005
COD					30	
BOD (with ATU)	<3		<5		<7	
DO percent saturation	>70		>50		>30	
Nitrogen (kjeldahl)	1		2		3	
Ammonia (as NH_4)	0.05		1	1.5	2	
Total coliforms/100 ml	50		5000		50 000	
Faecal coliforms/100 ml	20		2000		20 000	
Faecal streptococci/100 ml	20		1000		10 000	
Salmonella	absent in 51		absent in 11			4

Treatment types:

A1   Simple physical treatment and disinfection.
A2   Normal full physical and chemical treatment with disinfection.
A3   Intensive physical and chemical treatment with disinfection.

Mandatory levels 95% compliance, 5% not complying should not exceed 150% of mandatory level.

* EEC Directive 16/6/75.

3. Substances undesirable in excessive amounts—this group includes a
   wide variety of substances, some of which may be directly harmful in
   high concentrations, others which may produce taste and odour

TABLE 2.5. SOME EXAMPLES OF WHO GUIDELINES FOR DRINKING
WATER QUALITY*

Characteristic	Action level	
Arsenic	0.05	mg/l
Cadmium	0.005	„
Chromium	0.05	„
Cyanide	0.1	„
Fluoride	1.5	„
Lead	0.05	„
Mercury	0.001	„
Nickel	0.1	„
Nitrate and nitrite nitrogen	10	„
Nitrite nitrogen	1.0	„
Selenium	0.01	„
Chloride	250	„
Sulphate	400	„
Hardness as $CaCO_3$	500	„
Total dissolved solids	1000	„
Aluminium	0.2	„
Copper	1.0	„
Iron	0.3	„
Manganese	0.1	„
Sodium	200	„
Zinc	5.0	„
Chlorophenols	0.1	μg/l
Chloroform	30	„
DDT	1.0	„
Heptachlor	30	„
Lindane	3.0	„
Monochlorobenzene	3.0	„
1,4-dichlorobenzene	0.1	„
2,4-D	100	„
Gross alpha activity	0.1	Bq/l
Gross beta activity	1.0	„
Colour	15	TCU
Turbidity	5	NTU
Taste	not objectionable to 90% of consumers	
pH	6.5 to 8.5	
Coliforms	absent in 100 ml	

* *Guidelines for Drinking Water Quality*, WHO, Geneva, 1982.

problems and others may not be directly troublesome in themselves but are indicators of pollution. Constituents in this group include: nitrate, fluoride, phenol, iron and manganese, chloride, TOC.

4. Toxic substances—a wide range of inorganic and organic chemicals can have toxic effects on man; the severity of the effects depending for a particular material on the dose received, period of exposure and other environmental factors. The main concern in water supply is with the possible long-term effects of chronic exposure to low levels of toxic compounds. The establishment of allowable concentrations in these circumstances is often difficult so that large factors of safety are usually employed. Constituents which may be considered as toxic include, arsenic, cyanide, lead, mercury, polycyclic aromatic hydrocarbons, organochlorine and organophosphorus compounds.

5. Microbiological parameters—in most parts of the world these parameters are by far the most significant in determining water quality for potable supply. Standards for microbiological quality are essentially based on the need to ensure that bacteria indicative of pollution by human wastes are absent.

Table 2.4 gives examples of guide and mandatory limits used in EEC countries for raw waters intended for potable supply. It will be noted that for some parameters the levels permitted depend upon the type of water treatment to be provided. The World Health Organization has recently replaced its Drinking Water Standards by new Guidelines for Drinking Water Quality. In these guidelines the term Action Level is used to indicate the level above which the presence of any constituent should be investigated with a view to taking effective remedial action, Table 2.5 gives examples of these recommendations.

Further Reading

HOLDEN, W. S. (ed.), *Water Treatment and Examination*, Churchill, London, 1970.
SAWYER, C. N. and MCCARTY, P. L., *Environmental Chemistry for Engineering*, 3rd Ed., McGraw-Hill, New York, 1978.

CHAPTER 3

Sampling and Analysis

To obtain a true indication of the nature of a water or wastewater it is first necessary to ensure that the sample is actually representative of the source. Having satisfied this requirement, the appropriate analyses must be carried out using standard procedures so that results obtained by different analysts can be directly compared.

3.1. Sampling

The collection of a representative sample from a source of uniform quality poses few problems and a single grab sample will be satisfactory. A grab sample will also be sufficient if the purpose of sampling is simply to provide a spot check to see whether particular limits have been complied with. However, most raw waters and wastewaters are highly variable in both quality and quantity so that a grab sample is unlikely to provide a meaningful picture of the nature of the source. This point is illustrated in Fig. 3.1 which shows typical flow and strength variations in a sewer. To obtain an accurate assessment in this situation it is necessary to produce a composite sample by collecting individual samples at known time intervals throughout the period and measuring the flow at the same time. By bulking the individual samples in proportion to the appropriate flows an integrated composite sample is obtained. Similar procedures are often necessary when sampling streams and rivers and with large channel sections it may be desirable to sample at several points across the section and at several depths. Various automatic devices are available to collect composite samples and these may operate on either a time basis or on a flow-proportional basis. Sampling of industrial wastewater discharges may be even more difficult since they are often intermittent in nature. In these circumstances it is

19

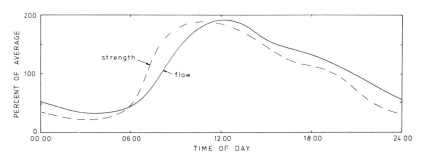

FIG. 3.1. Typical flow and strength variations in a sewer during dry weather.

important that the nature of the operations producing the discharge is fully understood so that an appropriate sampling programme can be drawn up to obtain a true picture of the discharge.

When designing a sampling programme it is vital that the objective of the exercise be clearly specified, e.g. to estimate maximum or mean concentrations, to detect changes or trends, to estimate percentiles or to provide a basis for industrial effluent charges. The degree of uncertainty which can be tolerated in the answer must also be specified and it is also necessary to bear in mind the resources available for sampling and analysis. For example, it may be found that to reduce the uncertainty of the results by a few percent might require twice the number of samples thus making the whole exercise uneconomic. It is therefore important to set a realistic level for the uncertainty of the results, based on the intended use. Ideally, all analyses should be carried out on the sample immediately after collection and certainly the quicker the analysis can be done the more likely it is that the results will be a true assessment of the actual nature of the liquid *in situ*. With characteristics which are likely to be unstable such as dissolved gases, oxidizable or reducible constituents, etc., the analyses must be carried out in the field or the sample must be suitably treated to fix the concentrations of unstable materials. Changes in the composition of a sample with time can be retarded by storage at low temperature (4°C) and the exclusion of light is also advisable. The more polluted a sample is, the shorter the time which can be allowed between sampling and analysis if significant errors are to be avoided.

3.2. Analytical Methods

Common analyses in the field of water quality control are usually based on relatively straightforward analytical principles. Quantitative analysis may be carried out by gravimetric, volumetric or colorimetric methods. Certain constituents may be determined by various types of electrode and there is increasing interest in the development of automated techniques for continuous monitoring of important parameters. It must be appreciated that, because of the low concentrations of impurities in water, laboratory work is often of a microanalytical nature requiring the use of careful procedures. Microbiological analysis is described in Chapter 4.

Gravimetric Analysis

This form of analysis depends upon weighing solids obtained from the sample by evaporation, filtration or precipitation. Because of the small weights involved, a balance accurate to 0.0001 g is required together with a drying oven to remove all moisture from the sample. Gravimetric analysis is thus not suited for field testing. Its main uses are:

1. Total and volatile solids. A known volume of sample in a preweighed nickel dish is evaporated to dryness on a water bath, dried at 103°C for wastewaters and 180°C for potable waters, and weighed. The increase in weight is due to the total solids. The loss in weight on firing at 500°C represents the volatile solids.

2. Suspended solids (SS). A known volume of sample is filtered under vacuum through a preweighed glass-fibre paper (Whatman GF/C) with a pore size of 0.45 μm. Total SS are given by the increase in weight after drying at 103°C and volatile SS are those lost on firing at 500°C.

3. Sulphate. For concentrations above 10 mg/l it is possible to determine sulphate by precipitating barium sulphate after the addition of barium chloride. The precipitate is filtered out of the sample, dried and weighed.

Volumetric Analysis

Many determinations in water quality control can be rapidly and accurately carried out by volumetric analysis, a technique which depends on the measurement of volumes of liquid reagent of known strength. The requirements for volumetric analysis are relatively simple:

1. A pipette to transfer a known volume of the sample to a conical flask.
2. A standard solution of the appropriate reagent. It is often convenient to make the strength of the standard solution such that 1 ml of the solution is chemically equivalent to 1 mg of the substance under analysis.
3. An indicator to show when the end point of the reaction has been reached. Various types of indicator are available, e.g. electrometric, acid-base, precipitation, adsorption and oxidation-reduction.
4. A graduated burette for accurate measurement of the volume of standard solution necessary to reach the end point.

An example of the use of volumetric analysis is found in the determination of alkalinity and acidity. Only with strong acids and strong bases does neutralization occur at pH 7; with all other combinations the neutralization points occur at pH 8.2 and pH 4.5. The indicators normally adopted for acidity and alkalinity are thus phenolphthalein (pink above pH 8.2, colourless below pH 8.2) and screened methyl orange (green above pH 4.5 and purple below pH 4.5). For the most accurate determinations in acid-base titrations a pH meter may be used for direct indication of the end point. Using N/50 standard solutions for alkalinity, acidity and also hardness determinations, 1 ml of titrant solution $\equiv$ 1 mg $CaCO_3$, the common denominator in which all these parameters are conventionally expressed.

Volumetric analysis can be useful in establishing the particular forms of alkalinity present in a sample. Neutralization of OH^- is complete at pH 8.2, whereas neutralization of $CO_3^=$ is only half completed at pH 8.2 and not fully completed until pH 4.5 is reached:

$$CO_3^= + H^+ \rightarrow HCO_3^- + H^+ \rightarrow H_2CO_3$$

Examination of the titration results will indicate the composition of the alkalinity present if it is assumed that HCO_3^- and OH^- cannot exist together. This is not strictly true and for accurate work more detailed considerations must be made. Reference to Fig. 3.2 shows that the following possibilities exist:

(a) OH^- alone will give an initial pH of about 10 and in this case, OH^- alkalinity = caustic alkalinity = total alkalinity.

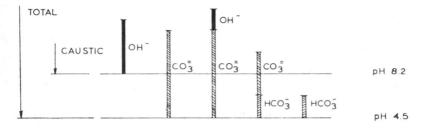

Fig. 3.2. Forms of alkalinity.

(b) $CO_3^=$ alone will give an initial pH of about 9.5,
    $CO_3^=$ alkalinity = 2 × caustic alkalinity = total alkalinity.

(c) OH^- and $CO_3^=$ together will give an initial pH about 10,
    $CO_3^=$ alkalinity = 2 × titration from pH 8.2 to pH 4.5,
    OH^- alkalinity = total alkalinity − $CO_3^=$ alkalinity.

(d) $CO_3^=$ and HCO_3^- together will give an initial pH > 8.2 and < 10.5,
    $CO_3^=$ alkalinity = 2 × caustic alkalinity,
    HCO_3^- alkalinity = total alkalinity − $CO_3^=$ alkalinity.

(e) HCO_3^- alone will give initial pH < 8.2,
    HCO_3^- alkalinity = total alkalinity.

Other common uses of volumetric analysis are in the determination of chloride (silver nitrate with potassium chromate precipitation indicator), the Winkler dissolved oxygen determination (sodium thiosulphate with starch adsorption indicator) and in the COD determination (ferrous ammonium sulphate with Ferroin ORP indicator).

Colorimetric Analysis

When dealing with low concentrations, colorimetric analyses are often particularly appropriate and there are many determinations in water quality control which can be quickly and easily carried out by this form of analysis.

To be of quantitative use a colorimetric method must be based on the formation of a completely soluble product with a stable colour. The coloured solution must conform with the following relationships:

(i) Beer's Law:
Light absorption increases exponentially with the concentration of the absorbing solution.

(ii) Lambert's Law:
Light absorption increases exponentially with the length of the light path.

These laws apply to all homogeneous solutions and can be combined as

$$OD = \log \frac{I_0}{I} = abc \qquad (3.1)$$

where OD = optical density,
I_0 = intensity of light entering sample,
I = intensity of light leaving sample,
a = constant characteristic of particular solution,
b = length of light path in solution,
c = concentration of absorbing substance in solution.

The colour produced may be measured by a variety of methods.

1. *Visual methods*

(a) Comparison tubes (Nessler tubes). A standard range of concentrations of the substance under analysis is prepared and the appropriate reagent added. The unknown sample is treated in the same manner and matched to the standards by looking down through the solutions on to a white base. The procedure is time consuming since the standards fade and must be re-made at intervals.

(b) Colour discs. In this case the standards are in the form of a series of suitably coloured glass filters through which a standard depth of distilled water or sample without colour-forming reagents is viewed. The sample in a similar tube is compared with the colour disc and the best visual match selected.

Both of these methods are dependent upon somewhat subjective judgements so that reproducibility between different analysts may not be good. The colour disc method is very convenient for field use and a wide range of discs and prepacked reagents are available.

2. *Instrumental methods*

    (a) Absorptiometer or colorimeter. This type of instrument comprises a glass sample cell through which a beam of light from a low-voltage lamp is passed. Light emerging from the sample is detected by a photo-electric cell whose output is displayed on a meter. The sensitivity is enhanced by inserting in the light path a colour filter complementary to the solution colour and the range of measurement can be extended by using sample cells of different length.

    (b) Spectrophotometer. This is a more accurate type of instrument using the same basic principle as an absorptiometer but with a prism being employed to give monochromatic light of the desired wavelength. The sensitivity is thus increased and on the more expensive instruments measurements can be undertaken in the infra-red and ultra-violet regions as well as in the visible light wavebands.

With both types of instrument a blank of the sample without the last colour-forming reagent is used to set the zero optical density position. The treated sample is then placed in the light path and the optical density noted. A calibration curve must be obtained by determining the optical density of a series of known standards at the optimum wavelength, obtained from analytical reference books or by experiment. In any form of colorimetric analysis it is important to ensure that full colour development has taken place before measurements are made and that any suspended matter in the sample has been removed. Suspended matter will of course prevent the transmission of light through a sample so that its presence will reduce the sensitivity of the determination and lead to erroneous results unless the blank has the identical concentration of suspended solids. In passing it may be noted that turbidity in samples is usually determined by nephelometry, a photo-electric technique which measures the scattering of light by colloidal particles.

Electrode Techniques

The measurement of such parameters as pH and ORP by electrodes has been widespread for many years and the technology of such electrodes is thus well established. pH is measured by the potential produced by a glass electrode—an electrode with a special sensitive glass area and an acid electrolyte, used in conjunction with a standard calomel reference electrode. The output from the pH electrode is fed to an amplifier and then to a meter or digital display. A wide range of pH electrodes is available including combined glass and reference units and special rugged units for field use. ORP is measured using a redox probe with a platinum electrode in conjunction with a calomel reference electrode.

More recent developments in electrode technology have resulted in the availability of a widening range of other electrodes some of which are extremely useful in water quality control. Probably the most useful of these new electrodes is the oxygen electrode. These DO electrodes come in a number of configurations using lead/silver, carbon/silver or gold/silver cells, sheathed with a polythene film. Polythene is permeable to oxygen so that oxygen in the sample enters the cell and alters its electrical output in proportion to the oxygen concentration. An increasing number of specific-ion electrodes for determinations such as NH_4^+, $NO_3^=$, Ca^{++}, Na^+, Cl^-, Br^-, F^-, etc., are now available. These electrodes permit rapid measurements down to very low concentrations but they are relatively costly and in some cases their stability is not particularly high.

3.3. Automated Analysis and Remote Monitoring

In laboratories with a large number of samples for analysis there is considerable use of automated techniques to speed up the work and reduce staffing requirements. Many of these automated analytical techniques utilize colorimetric determinations with an automatic sampler feeding discrete samples through the necessary reagent addition and colour-development stages to a spectrophotometer, the output from which can be recorded on strip charts or in computer-compatible format. Equipment of this nature operates in a normal laboratory environment and skilled human supervision is available, at least during usual working hours. In contrast with this situation there is a need for the operation of continuous remote monitoring installations, particularly on raw water sources, to provide early warning of any changes in water quality. For this purpose the instrumentation may have to be sited in a somewhat inhospitable environment and will be required to operate for extended periods without attention other than routine maintenance and calibration. With these constraints it is usually necessary to restrict the parameters monitored to those which can be determined by simple colorimetric or electrode techniques and which will give a continuous record of water quality. Most remote monitoring stations are thus equipped with sensors for parameters such as, pH, temperature, conductivity, turbidity, DO and possibly ammonia and nitrate nitrogen. The installation of dual sensors and some means of regular automatic calibration and cleaning are necessary if the results from the station are to be reliable. Data collected by remote stations, which should ideally be situated at a flow gauging station, may be transmitted back to a suitable base control for data logging with provision for generating an alarm state if predetermined levels for any of the parameters are passed. It is not of course possible to monitor a water for all possible contaminants but it would be desirable if some form of monitor were available to indicate the presence of potentially toxic chemicals in a raw water used for potable supply. A number of such early warning devices have been developed based on the effect of toxic constituents on fish or nitrifying bacteria kept in a test chamber through which the water flows. A change in the activity of the fish or the nitrifying performance of the bacteria indicates the possible presence of a toxic pollutant although detailed chemical analysis will then be necessary to identify the offending constituent.

Further Reading

BRIGGS, R., Instrumentation for monitoring water quality. *Wat. Treat. Exam.* **24**, 1975, 23.

ELLIS, J. C. and LACEY, R. F., Sampling: defining the task and planning the scheme. *Wat. Pollut. Control*, **79**, 1980, 452.

GARRY, J. A., MOORE, C. J. and HOOPER, B. D., Sewage treatment works effluent sampling—have our means any meaning? *Wat. Pollut. Control*, **80**, 1981, 481.

HEY, A. E., Continuous monitoring for sewage-treatment processes. *Wat. Pollut. Control*, **79**, 1980, 477.

HINGE, D. C., Experiences in the continuous monitoring of river water quality. *J. Instn Wat. Engrs Scits*, **34**, 1980, 546.

MORLEY, P. J. and COPE, J., Water quality monitoring. In *Developments in Water Treatment*, **2** (Ed. Lewis, W. M.), Applied Science Publishers Ltd., London, 1980, 189.

SAWYER, C. N. and MCCARTY, P. L., *Environmental Chemistry for Engineering*, 3rd edn, McGraw-Hill, New York, 1978.

SCHOFIELD, T., Sampling of water and wastewater: practical aspects of sample collection. *Wat. Pollut. Control*, **79**, 1980, 477.

Standard Methods of Analysis

AMERICAN PUBLIC HEALTH ASSOCIATION, *Standard Methods for the Examination of Water and Wastewater*, 15th edn, APHA, New York, 1980.

DEPARTMENT OF THE ENVIRONMENT/NATIONAL WATER COUNCIL, *Methods for the Examination of Waters and Associated Materials* (separate booklets for each determination), HMSO, London, various dates.

CHAPTER 4

Aquatic Microbiology and Ecology

A FEATURE of most natural waters is that they contain a wide variety of micro-organisms forming a balanced ecological system. The types and numbers of the various groups of micro-organisms present are related to water quality and other environmental factors. In the treatment of organic wastewaters, micro-organisms play an important role and most of the species found in water and wastewater are harmless to man. However, a number of micro-organisms are responsible for a variety of diseases and their presence in water poses a health problem. It is therefore necessary to develop an understanding of the basic principles of microbiology and thus gain an appreciation of the role of micro-organisms in water quality control.

4.1. Types of Metabolism

Virtually all micro-organisms require a moist environment for active growth but apart from this common feature many different types of metabolism are found. A basic classification can be made in relation to whether or not an organism requires an external source of organic matter.

Autotrophic organisms are capable of synthesizing their organic requirements from inorganic matter and can thus grow independently of external organics. Two methods may be employed to achieve this end:

1. Photosynthesis—many plants can utilize inorganic carbon and ultra-violet radiation to produce organic matter and oxygen

$$6CO_2 + 6H_2O \xrightarrow{\text{light}} C_6H_{12}O_6 + 6O_2$$

2. Chemosynthesis—chemical energy of inorganic compounds is utilised to provide the energy for synthesis of organics

$$2NH_3 + 3O_2 \rightarrow 2HNO_2 + 2H_2O + energy$$

Heterotrophic organisms require an external source of organic matter and are of three main types:

1. Saprophobes which obtain soluble organic matter directly from the surrounding environment or by extracellular digestion of insoluble compounds. Food requirements can range from a simple organic carbon compound to a number of complex carbon and nitrogen compounds together with additional growth factors.
2. Phagotrophes, sometimes termed holozoic forms, can utilize solid organic particles.
3. Paratrophes obtain organic matter from the tissues of other living organisms and are thus parasitic.

Organisms differ in their requirements for oxygen, aerobes require the presence of free oxygen, whereas anaerobes exist in the absence of free oxygen. Facultative forms have a preference for one form of oxygen environment but can live in the other if necessary. In terms of temperature requirements there are three main types of organisms; psychrophilic which live at a temperature close to $0°C$; mesophilic, by far the most common, living within the range $15–40°C$ and thermophilic in the range $50–70°C$. In practice there is some overlap between these temperature ranges so that some organisms will be found growing actively at any temperature between 0 and $70°C$.

4.2. Types of Micro-organism

By definition, micro-organisms are those organisms too small to be seen by the naked eye and there are large numbers of aquatic organisms in this category. With higher organisms it is convenient to identify them as plants or animals. Plants have rigid cell walls, are photosynthetic and do not move independently. Animals have flexible cell walls, require organic food and are capable of independent movement. The application of such differentiation to micro-organisms is difficult because of the simple structures of their cells and it has become convention to term all micro-organisms protists. The protists can themselves be divided into two types:

Procaryotes—small ($< 5\,\mu$m) simple cell structures with rudimentary nucleus and one chromosome. Reproduction is normally by binary fission. Bacteria, actinomycetes and the blue-green algae are included in this group.

Eucaryotes—larger ($> 20\,\mu$m) cells with a more complex structure and containing many chromosomes. Reproduction may be asexual or sexual and quite complex life cycles may be found. This class of micro-organisms includes fungi, most algae and the protozoa.

There is a further group of micro-organisms, the viruses, which do not readily fit into either of the above classes and which are thus considered separately.

Viruses

Viruses are the simplest form of organism ranging in size from about 0.01 to 0.3 μm and they consist essentially of nucleic acid and protein. They are all parasitic and cannot grow outside another living organism. All are highly specific both as regards the host organism and the disease which they produce. Human viral diseases include, smallpox, infectious hepatitis, yellow fever, poliomyelitis and a variety of gastro-intestinal diseases. Because of the inability of viruses to grow outside a suitable host they are on the borderline between living matter and inanimate chemicals. Identification and enumeration of viruses requires special apparatus and techniques. Sewage effluents normally contain significant numbers of viruses and they also present in most surface waters subject to pollution. Because of their small size their removal in conventional water-treatment processes cannot be certain although the normal disinfection processes will usually inactivate viruses.

Bacteria

Bacteria are the basic units of plant life, they are single-cell organisms which utilise soluble food and may operate either as autotrophes or as heterotrophes. They range in size from 0.5 to 5 μm and have the basic features shown in Fig. 4.1. Reproduction is by binary fission and the generation time may be as short as 20 min in favourable conditions with some species. Some bacteria can form resistant spores which can remain dormant for long periods in unsuitable environmental conditions but which can be reactivated on the return of suitable conditions. Most bacteria prefer more or less neutral pH conditions although some species can exist in a highly acid environment. Bacteria play a vital role in the natural stabilization processes and are widely utilized for the treatment of organic wastewaters. There are some 1500 known species which are classified in relation to criteria such as: size, shape and grouping of cells; colony characteristics; staining behaviour; growth requirements; motility, specific chemical reactions; etc. Aerobic, anaerobic and facultative forms are found.

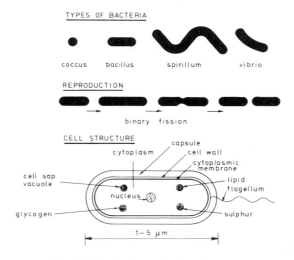

FIG. 4.1. Basic characteristics of bacteria.

Fungi

Fungi are aerobic multicellular plants which are more tolerant of acid conditions and a drier environment than bacteria. They utilize much the same food sources as the bacteria in chemosynthetic reactions but, because their protein content is somewhat lower than the bacteria, their nitrogen requirement is less. Fungi form rather less cellular matter than bacteria from the same amount of food. They are capable of degrading highly complex organic compounds and some are pathogenic in man. Over 100 000 species of fungi exist and they usually have a complex structure formed of a branched mass of thread-like hyphae (Fig. 4.2). They have four or five distinct life phases with reproduction by asexual spores or seeds. Fungi occur in polluted waters and in biological treatment plants, particularly in conditions with high C:N ratios. They can be responsible for tastes and odours in water supplies.

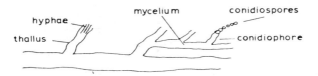

FIG. 4.2. Main features of fungi.

Actinomycetes

The actinomycetes are similar to fungi in appearance with a filamentous structure but with a cell size close to that of bacteria. They occur widely in soil and water and nearly all are aerobic. Their significance in water is mainly due to the taste and odour problems which often result from their presence.

Algae

Algae are all photosynthetic plants, mostly multicellular although some types are unicellular. The majority of freshwater forms utilize the pigment chlorophyll and they act as the main producers of organic matter in an aquatic environment. Inorganic compounds such as carbon dioxide, ammonia, nitrate and phosphate provide the food source to synthesize new algal cells and to produce oxygen. In the absence of sunlight, algae operate on a chemosynthetic basis and consume oxygen so that in a water containing algae there will be a diurnal variation in DO levels; supersaturation may occur during the day with significant oxygen depletion occuring at night. A large number of algae are found in freshwater and a variety of classification systems exists. Algae may be green, blue-green, brown or yellow depending upon the proportions of particular pigments. They occur as single cells which may be motile with the aid of flagella or non motile, or as multicellular filamentous forms (Fig. 4.3). Algae and bacteria growing in the same solution do not compete for food but have a symbiotic relationship (Fig. 4.4) in which the algae utilize the end products of bacterial decomposition of organic matter and produce oxygen to maintain an

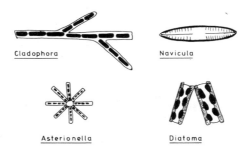

Cladophora　　Navicula

Asterionella　　Diatoma

Fig. 4.3. Some typical algae.

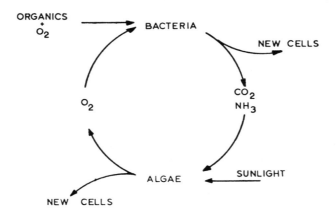

Fig. 4.4. Symbiotic relationship between bacteria and algae.

aerobic system. In the absence of organic input, algal growth depends upon the mineral content of the water so that in a hard water algae obtain CO_2 from bicarbonates, reducing the hardness and usually increasing the pH. Algae are important in water because of their effect on DO levels and because some species can produce severe taste and odour problems. In a few instances, cattle have died after drinking water containing algal products. Since some species of algae can utilize atmospheric nitrogen the critical element controlling the growth of algae is usually phosphorus.

Protozoa

The protozoa are unicellular organisms 10–100 μm in length which reproduce by binary fission. Most are aerobic heterotrophes and often utilize bacterial cells as their main food source. They cannot synthesize all the necessary growth factors and rely on the bacteria to provide these items. The protozoa are widespread in soil and water and may sometimes play an important role in biological waste-treatment processes. There are four main types of protozoa (Fig. 4.5): *Sarcodina*—amoeboid flexible cell structure with movement by means of extruded pseudopod (false foot); *Mastigophora*—utilize flagella for motility; *Ciliata*—motility and food gathering by means of cilia (hair-like feelers), may be free swimming or stalked; *Sporozoa*—non-motile spore-forming parasites not found in water.

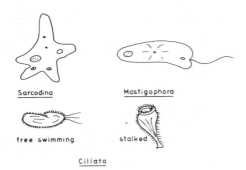

Sarcodina   Mastigophora

free swimming   stalked

Ciliata

Fig. 4.5. Some typical protozoa.

Higher Forms of Life

As well as the micro-organisms more complex macro-organisms, many visible to the naked eye, are found in natural waters. These include, rotifers which are multicellular animals with a flexuous body and cilia on the head to catch food and provide motility, and crustaceans which are hard-shelled multicellular animals. Both of these groups provide important food supplies for fish and are normally only found in good-quality waters since they are sensitive to many pollutants and to low DO levels. Worms and insect larvae are found in bottom deposits and in some biological treatment processes and they can often metabolize complex organics not readily broken down by other organisms.

4.3. Nomenclature

The nomenclature used in biology appears complex to the non-biologist but is very necessary because of the multiplicity of organisms which exists. A specific type of organism is denoted by its specific name and a collection of similar species is given a generic name, e.g.

Salmonella paratyphi, a member of the *Salmonella* genus—the bacteria specifically responsible for paratyphoid.

Entamoeba histolytica, the amoebic protozoan responsible for amoebic dysentery.

Drifting aquatic micro-organisms are collectively termed phytoplankton if of plant origin and zooplankton when of animal origin. Free-swimming groups are called nekton, surface-swimming types are called neuston and bottom-living groups are referred to as benthos.

4.4. Microbiological Examination

Because of their small size, observation of micro-organisms with the naked eye is impossible and in the case of the simpler micro-organisms their physical features do not provide positive identification. With bacteria it is necessary to utilize their biochemical or metabolic properties to aid identification of individual species. Living specimens are often difficult to observe because they usually have little colour and thus do not stand out against the liquid background. Staining of dead specimens is useful in some cases but the staining and mounting technique may itself alter some of the cell characteristics.

An optical microscope has a maximum magnification of about 1000 × with a limit of resolution of about 0.2 μm. Hence most viruses will be invisible with an optical microscope and only limited detail of the structure of a bacterial cell can be observed. For more detailed examination it is necessary to use an electron microscope which can provide magnifications of around 50 000 × with a limit of resolution of about 0.01 nm.

Study of living specimens, which is only possible under an optical microscope, is necessary to determine whether or not an organism is motile. A hanging drop slide must be used in this case and it is important to differentiate between Brownian movement, a random vibrational motion common to all colloids, and true motility which is a rapid shooting or wriggling movement.

In many situations it is necessary to assess the number of micro-organisms present in a water sample. With the larger micro-organisms like algae, estimation of numbers and actual identification of species can be achieved using a special microscope slide containing a depression of known volume. Numbers are obtained by counting the appropriate micro-organisms in the chamber, aided by a grid etched onto the slide.

An estimate of the number of living bacteria (viable cell count) in a water sample may be obtained with a plate count using nutrient agar medium. A 1-ml sample of the water, diluted if necessary, is mixed with liquefied agar at 40°C in a petri dish. The agar sets to a jelly thus fixing the bacterial cells in position. The plate is then incubated under appropriate conditions (72 h at 22°C for natural water bacteria, 24 h at 37°C for bacteria originating from animals or man). At the end of incubation the individual bacteria will have produced colonies visible to the naked eye and the number of colonies is assumed to be a function of the viable cells in the original sample. In practice such plate counts do not give the total population of a sample, since no single medium and temperature combination will permit all bacteria to reproduce. However, viable counts at the two temperatures using a wide-spectrum medium do give an overall picture of the bacterial quality and pollution history of a sample.

To determine the presence of a particular genus or species of bacteria it is necessary to utilize its characteristic behaviour by supplying special selective media and/or incubation conditions which are suitable only for the bacteria under investigation. As is described in Chapter 5, many serious diseases are related to microbiological contamination of water, most of them due to pathogenic bacteria excreted by people suffering from or carrying the disease. Whilst it is possible to examine a water for the presence of a specific pathogen a more sensitive test employs an indicator organism *Escherichia coli* which is a normal inhabitant of the human intestine and is excreted in large numbers. Its presence in water thus indicates human excretal contamination and the sample is therefore potentially dangerous in that pathogenic faecal bacteria *might* also be present. Coliform bacteria in general have the ability to ferment lactose to produce acid and gas. Detection of coliforms can be achieved using a lactose medium (MacConkey Broth) inoculated with serial dilutions of the sample. The appearance of acid and gas after 24 h at 37°C is taken as positive indication of the presence of coliform bacteria, results being expressed with the aid of statistical tables as most probable number (MPN)/100 ml. As a confirmatory test for *Escherichia coli*, positive tubes are subcultured in fresh medium for 24 h at 44°C under which conditions only *E. coli* will grow to give acid and gas.

An alternative technique for bacteriological analysis, which is now popular, uses special membrane filter papers with a pore size such that bacteria can be separated from suspension. The bacteria retained on the paper are then placed in contact with an absorbent pad containing the appropriate nutrient medium in a small plastic petri dish and incubated. Identification of a particular species of bacteria is made on the basis of the type of nutrient provided and often on the appearance (colour, sheen) of the colonies formed. Counting the colonies provides the necessary quantitative information. The membrane filter technique is very convenient for field testing although the cost of the materials is likely to be higher than for the conventional methods of analysis unless the filters are washed for re-use.

4.5. Ecological Principles

In all communities of living organisms the various forms of life are interdependent to a greater or lesser extent. This interdependence is

essentially nutritional, described as a trophic relationship, and is exemp-
lified by the cycle of organic productivity and the carbon and nitrogen
cycles. A biological community and the environment in which it is found
form an ecosystem and the science of such systems is known as ecology.

The autotrophes in an ecosystem, i.e. green plants and some bacteria, are
termed *producers* since they synthesize organic matter from inorganic
constituents. Heterotrophic animals are known as *consumers* since they
require ready-made organic food and may be subdivided into herbivores
(plant eaters) and carnivores (meat eaters). Heterotrophic plants are termed
decomposers since they break down the organic matter in dead plants and
animals and in animal excreta. Some of the products of decomposition are
utilized for their own growth and energy requirements, but others are
released as simple inorganic compounds suitable for plant uptake. Figure
4.6 shows a simplified version of the carbon cycle. Solar radiation provides
the only external energy source and permits the synthesis of carbohydrates
and other organic products which are then transferred to the heterotrophic
phase of the cycle along with oxygen resulting from photosynthesis. In
exchange, carbon dioxide, water and inorganic salts resulting from the
activities of animals and bacteria are returned to the autotrophes. It should
be noted that whilst carbon follows a cyclical path in such a system, energy
flow is one-way only. It is important to remember that a continual energy
input is thus necessary to allow the system to function. The loss of some part
of the energy input to heat and entropy which inevitably occurs in biological
systems can be considered as analogous to friction losses in a mechanical

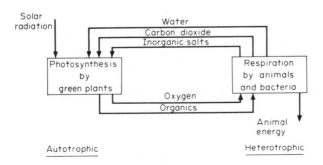

FIG. 4.6. The carbon cycle.

system. In fact the efficiency in terms of energy conversion of biological systems is very low and the further away an organism is from the original energy input the lower will be the proportion of that energy available to the particular organism.

In an aquatic environment the interdependence of organisms takes the form of a complex food web within which are many food chains with successive links being composed of different species in a predator–prey relationship with adjacent links. Thus a typical food chain for a river would be

algae → rotifer → mayfly → minnow → pike

The successive links in the food chain contain fewer but larger individual organisms and the community can be pictured in the form of an Eltonian pyramid of numbers (Fig. 4.7). Each level in the pyramid is known as a trophic level. Organisms occupying the same level compete for a common food, but those on a higher level are predatory on the lower level. Under natural conditions such an ecosystem can remain dynamically balanced over long periods, but changes in water quality or other environmental factors can completely upset the balance. Toxic materials tend to give a particular percentage kill of the population regardless of the population density, whereas the effect of such factors as shortage of food is more likely

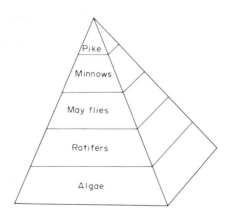

FIG. 4.7. An Eltonian pyramid of numbers.

to be significant with dense populations. A clean surface water will normally contain many different forms of life, but none will be dominant and the community will be well balanced. Serious organic pollution of the water would produce conditions unsuitable for most of the higher forms of life so that the community would become one of one or two simple life forms which would be present in very large numbers because of the absence of predators.

Because of the low efficiency of ecological processes the numbers of organisms at the first trophic level required to support an organism at the top of the pyramid becomes very large. As a result, food chains in nature do not contain more than five or possibly six trophic levels.

Further Reading

BELCHER, H. and SWALE, E., *A Beginner's Guide to the Freshwater Algae*, HMSO, London, 1976.

HAWKES, H. A., *The Ecology of Waste Water Treatment*, Pergamon Press, Oxford, 1963.

HOLDEN, W. S. (Ed.), *Water Treatment and Examination*, Churchill, London, 1970.

MACAN, T. T. and WORTHINGTON, E. B., *Life in Lakes and Rivers*, Fontana, London, 1972.

MARA, D. D., *Bacteriology for Sanitary Engineers*, Churchill Livingstone, Edinburgh, 1974.

McKINNEY, R. E., *Microbiology for Sanitary Engineers*, McGraw-Hill, New York, 1962.

MINISTRY OF HOUSING AND LOCAL GOVERNMENT, Reports on Public Health and Medical Subjects No. 71, *The Bacteriological Examination of Water Supplies*, 4th edn, HMSO, London, 1969.

MITCHELL, R., *Introduction to Environmental Microbiology*, Prentice Hall, Englewood Cliffs, 1974.

ODUM, E. P., *Fundamentals of Ecology*, 3rd edn, W. B. Saunders Company, Philadelphia, 1971.

CHAPTER 5

Water Quality and Health

BECAUSE of the essential role played by water in supporting human life it also has, if contaminated, great potential for transmitting a wide variety of diseases and illnesses. In the developed world water-related diseases are rare, due essentially to the presence of efficient water supply and wastewater disposal systems. However, in the developing world perhaps as many as 2000 million people are without safe water supply and adequate sanitation. As a result, the toll of water-related disease in these areas is frightening in its extent. A recent WHO survey has highlighted the following facts:

> Each day some 30 000 people die from water-related diseases. In developing countries 80 percent of all illness is water-related. A quarter of children born in developing countries will have died before the age of 5, the great majority from water-related disease.
>
> At any one time there are likely to be 400 million people suffering from gastroenteritis, 200 million with schistosomiasis, 160 million with malaria and 30 million with onchoceriasis. All of these diseases can be water-related although other environmental factors may also be important.

In the developed world there is concern about the possible long-term health hazards which may arise from the presence of trace concentrations of impurities in drinking water, particular attention being paid to potentially carcinogenic compounds. There are also several contaminants, which may be naturally occurring or man-made, having known effects on the health of consumers. It is therefore important that the relationships between water quality and health be fully appreciated by the engineers and scientists concerned with water quality control.

5.1. Characteristics of Diseases

Before considering the water-related diseases it is necessary to briefly outline the main features of communicable diseases.

All diseases require for their spread a source of infection, a transmission route, and the exposure of a susceptible living organism. Control of disease is thus based on curing sufferers, breaking the transmission route and protecting the susceptible population. Engineering measures in disease control are essentially concerned with breaking the transmission route and medical measures are concerned with the other two parts of the infection chain.

Contagious human diseases are those where the pathogen spends its life in man and can only live a short time in the unfavourable environment outside the body. This type of disease is thus transmitted by direct contact, droplet infection or similar means. With non-contagious diseases the pathogen spends part of its life cycle outside the human body so that direct contact is not of great significance. Non-contagious diseases may involve simple transmission routes with extracorporeal development of the infective organism taking place in soil or water. In many cases, however, more complex transmission routes occur with requirements for an intermediate host as part of the development of the parasite. It is thus important that control measures are developed in full knowledge of the transmission patterns of the particular disease. When a disease is always present in a population at a low level of incidence it is termed *endemic*. When a disease has widely varying levels of incidence the peak levels are called *epidemics* and world-wide outbreaks are termed *pandemics*.

5.2. Water-related Disease

There are about two dozen infectious diseases, shown in Table 5.1, the incidence of which can be influenced by water. These diseases may be due to viruses, bacteria, protozoa or worms and although their control and detection is based in part on the nature of the causative agent it is often more helpful to consider the water-related aspects of the spread of infection. In the context of diseases associated with water there has in the past been some confusion about the terminology applied. Bradley[1] has developed a

TABLE 5.1. MAIN WATER-RELATED DISEASES

Disease	Type of water relationship
Cholera Infectious hepatitis Leptospirosis Paratyphoid Tularaemia Typhoid	Waterborne
Amoebic dysentery Bacillary dysentery Gastroenteritis	Waterborne or Water-washed
Ascariasis Conjunctivitis Diarrhoeal diseases Leprosy Scabies Skin sepsis and ulcers Tinea Trachoma	Water-washed
Guinea worm Schistosomiasis	Water-based
Malaria Onchocerciasis Sleeping sickness Yellow fever	Water-related insect vector

more specific classification system for water-related diseases which differentiates between the various forms of infections and their transmission routes.

Waterborne Disease

The commonest form of water-related disease and certainly that which causes most harm on a global scale includes those diseases spread by the contamination of water by human faeces or urine. With this type of disease, infection occurs as shown in Fig. 5.1 when the pathogenic organism gains access to water which is then consumed by a person who does not have immunity to the disease. The majority of diseases in this category, cholera,

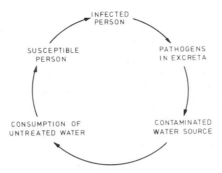

FIG. 5.1. The classical waterborne disease infection cycle.

typhoid, bacillary dysentry, etc., follow a classical faecal–oral transmission route and outbreaks are characterized by simultaneous illness amongst a number of people using the same source of water. It should be appreciated that although these diseases can be waterborne they can also be spread by any other route which permits direct ingestion of faecal matter from a person suffering from that disease. Poor personal hygiene of workers in food handling and preparation activities would provide an obvious infection route. The situation is further complicated in that some people may be carriers of diseases like typhoid so that although they exhibit no outward signs of the disease their excreta contain the pathogens. Screening for such a carrier state is often practised with potential employees in the water-supply industry.

There are other waterborne diseases in which the infection pattern is not so simple. Weil's disease (leptospirosis) is transmitted in the urine of infected rats and the causative organism is able to penetrate the skin so that external contact with contamined sewage or flood water can spread the disease.

Water-washed Disease

As described above, infections which are transmitted by the ingestion of faecally contaminated water can also be spread by more direct contact between faeces and mouth. In the case of poor hygiene, due to inadequate water supply for washing, the spread of infection may be reduced by providing additional water, the quality of which becomes a secondary consideration. Water in this context is a cleansing agent and since it is not ingested the normal quality requirements need not be paramount. Clearly, waterborne faecal–oral diseases may also be classified as water-washed diseases and many of the diarrhoeal infections in tropical climates behave as water-washed rather than waterborne diseases.

A second group of diseases can also be classified as water-washed. These include a number of skin and eye infections which whilst not normally fatal have a serious debilitating effect on sufferers. The diseases of this type include bacterial ulcers and scabies, and trachoma. They tend to be associated with hot dry climates and their incidence can be significantly reduced if ample water is available for personal washing. This second group of water-washed infections are not also waterborne as were the first group.

Water-based Disease

A number of diseases depend upon the pathogenic organism spending part of its life-cycle in water or in an intermediate host which lives in water. Thus infection of man cannot occur by immediate ingestion of or contact with the organism excreted by a sufferer. Most of the diseases in this class are caused by worms which infest the sufferer and produce eggs which are discharged in faeces or urine. Infection often occurs by penetration of the skin rather than by consumption of the water.

Schistosomiasis (also called bilharzia) is probably the most important example of this class of disease. The transmission pattern of schistosomiasis is relatively complex in comparison with waterborne diseases and is illustrated in Fig. 5.2. If a sufferer excretes into water, eggs from the worms hatch into larvae which can live for only 24 h unless they find a particular species of snail which acts as an intermediate host. The larvae then develop in a cyst in the snail's liver which after about 6 weeks bursts and releases minute free-swimming cercariae which can live in water for about 48 h. The cercariae have the property of being able to puncture the skin of man and other animals and they can then migrate through the body via skin, veins, lungs, arteries and liver in a period of around 8 weeks. The parasite then develops in the veins of the wall of the bladder, or of the intestine, into a worm which may live several years and which will discharge enormous

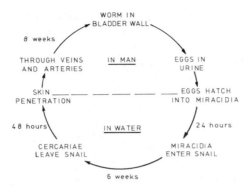

FIG. 5.2. The infection cycle for schistosomiasis.

number of eggs. It is unfortunate that schistosomiasis is often spread by irrigation schemes which, unless carefully designed and operated, tend to provide suitable habitats for the snail host as well as increasing the likelihood of contact with the water by agricultural workers. Control measures for schistosomiasis include, as well as the obvious prevention of excretal contamination of water, creation of conditions unfavourable to the presence of the snails, prevention of human contact with potentially contaminated water and the insertion of a 48-h delay period after removal of snails before access to the water is permitted. Unfortunately such measures are not easy to enforce and there may often be a conflict between the needs of schistosomiasis control and the desire to obtain agricultural and economic benefits from irrigation schemes in which some degree of water contact is inevitable.

Guinea worm is another water-based disease which is widespread in the tropics. In this case the intermediate host is cyclops, a small crustacean, and human infection occurs following the ingestion of water containing infected cyclops. Eggs are discharged when an ulcer on the skin of a sufferer bursts and they can remain viable in water for one or two weeks. If eggs are ingested by cyclops they develop into larval forms in a further 2 weeks. The larvae leave the ingested cyclops during human digestive processes and migrate through the tissues to the lower limbs of the body and eggs are discharged about 9 months later. The vector species of cyclops is prevalent in stagnant water with some organic content. Control of guinea worm, which can bring marked improvements in the health of the population, is essentially based on protection of water sources, particularly springs and wells. The provision of sloping hardstandings and parapets round water sources will effectively prevent the access of eggs to the water.

Water-related Insect Vectors

There are a number of diseases which are spread by insects which breed or feed near water so that their incidence can be related to the proximity of suitable water sources. Infection with these diseases is in no way connected with human consumption of or contact with the water. Mosquitoes which transmit malaria and a number of other diseases prefer shallow stagnant water in pools, around the edges of lakes and in water storage jars. It is therefore important to ensure that water supply and drainage works do not provide suitable mosquito habitats or, if this is unavoidable, mosquitoes should be prevented from gaining access by the provision of effective screens. *Simulium* flies which transmit onchocerciasis (river blindness) breed in turbulent waters associated with rapids, waterfalls, etc., or created by engineering structures like weirs, energy dissipators, etc. Control is usually by use of insecticides injected upstream of the point of turbulence.

5.3. Chemical-related Illness

There are many chemical compounds whose presence in water could be harmful or fatal to human life and it is necessary to consider two aspects of the problem in assessing potential hazards. An acute effect could be produced by the accidental discharge of sufficient toxic matter into a water source to produce more or less immediate symptoms in consumers. This form of contamination is fortunately rare and usually the contaminant would produce obvious effects in the water source such as fish kills, strong tastes and odours, etc., which would provide a warning even if the accident had not been reported to the authorities. A more insidious type of chemical contamination occurs when the contaminant produces a long-term hazard due to exposure to minute concentrations, perhaps over many years. In this situation the determination of allowable levels for the particular contaminants is extremely difficult since scientific evidence is likely to be very limited and difficult to interpret.

Probably one of the earliest chemical contamination problems arose from the use of lead piping and tanks in domestic plumbing. Soft acidic waters from upland catchments tend to be plumbosolvent so that significant amounts of lead can be dissolved in the water, particularly when standing overnight in service connections. Lead is a cumulative poison and current concern about lead in the environment has meant that allowable levels of lead are kept as low as possible. The EEC standards for drinking water specify a maximum level of 50 μg/l and this may be difficult to achieve in areas with plumbosolvent waters and lead plumbing. The addition of lime to the water reduces the dissolution of lead and fortunately the high cost of lead means that alternative materials such as copper and plastic have been used in domestic plumbing for many years.

Nitrate nitrogen occurs naturally in some groundwaters and is present in the runoff from agricultural land as well as in conventionally-treated sewage effluents. Although its presence in drinking water is not known to be harmful to children or adults it can be hazardous in the case of young babies up to the age of about 6 months. Up to this age babies do not have the normal bacterial flora in their intestines and are unable to deal with the nitrite produced by reduction of nitrate in the stomach. If a baby is bottle-fed with milk made from water containing more than 10–20 mg/l of nitrate nitrogen the possibility of methaemoglobinemia exists since nitrite absorbed in the blood prevents oxygen transport.

Fluoride is a natural constituent of some waters and it has been shown to have an inhibiting effect on tooth decay, particularly in children. As a result, some health authorities recommend fluoridation of water to a level of 1 mg/l. This recommendation is often vociferously opposed by those objecting to mass medication. At levels of fluoride above 1.5 mg/l yellow stains on the teeth may appear and at much higher levels bone fluorosis may occur.

Hardness in water may have some effect in reducing certain types of heart disease and thus softening a water may have a detrimental health effect. In addition some softening processes increase the sodium content of the water and this can be undesirable for some heart and kidney complaints.

New and sophisticated analytical techniques which have recently become available have enabled the determination of many trace organic compounds in water whose presence was previously unknown. The detection in raw waters of polyaromatic hydrocarbons (PAH), trihalomethanes (THM), organochlorine compounds (OCl) and organophosphorus compounds (OP) has caused some concern since substances of this type may be carcinogenic. There are thus possible long-term health hazards arising from their presence in drinking water although at the levels currently found, there is no conclusive evidence of their carcinogenicity.

Reference

1. BRADLEY, D. J., Health aspects of water supplies in tropical countries. In *Water, Wastes and Health in Hot Climates* (Ed. Feachem, R. G., McGarry, M. and Mara, D. D.), Wiley, Chichester, 1977.

Further Reading

BRITTON, A. and RICHARDS, W. W., Factors influencing plumbosolvency in Scotland. *J. Instn Wat. Engrs Scits*, **35**, 1981, 349.

FIELDING, M. and PACKHAM, R. F., Organic compounds in drinking water and public health. *J. Instn Wat. Engrs Scits*, **31**, 1977, 353.

INTERNATIONAL WATER SUPPLY ASSOCIATION, Nitrates in water supplies, *Aqua*, 1974, No. 1, 5.

WORLD HEALTH ORGANIZATION, *Health Hazards of the Human Environment*, WHO, Geneva, 1972.

WORLD HEALTH ORGANIZATION, *Epidemiology and Control of Schistosomiasis*, Technical Report 643, WHO, Geneva, 1980.

CHAPTER 6

Biological Oxidation of Organic Matter

M ANY of the problems associated with water quality control are due to the presence of organic matter from natural sources or in the form of wastewater discharges. This organic matter is normally stabilized biologically and the micro-organisms involved utilize either aerobic or anaerobic oxidation systems.

In the presence of oxygen, aerobic oxidation takes place, part of the organic matter being synthesized to form new micro-organisms and the remainder being converted to relatively stable end products as shown in Fig. 6.1. In the absence of oxygen, anaerobic oxidation will produce new cells and unstable end products such as organic acids, alcohols, ketones, methane.

The methane-producing system which is the commonest in waste treatment takes place in two stages; in the first stage acid-forming micro-organisms convert the organic matter into new cells and organic acids and alcohols. A second group of micro-organisms, the methane bacteria, then continue the oxidation again utilizing part of the organic matter to synthesize new cells and converting the remainder to methane, carbon dioxide and hydrogen sulphide. The anaerobic reaction is much slower than the aerobic process and is inefficient as regards energy conversion, aerobic oxidation of, say, glucose yielding about thirty times the energy released by its oxidation anaerobically.

6.1. Nature of Organic Matter

There are three main types of organic matter in water quality control:
1. Carbohydrates (CHO) containing carbon, hydrogen and oxygen. Typical examples are sugars, e.g. glucose $C_6H_{12}O_6$, starch, cellulose.

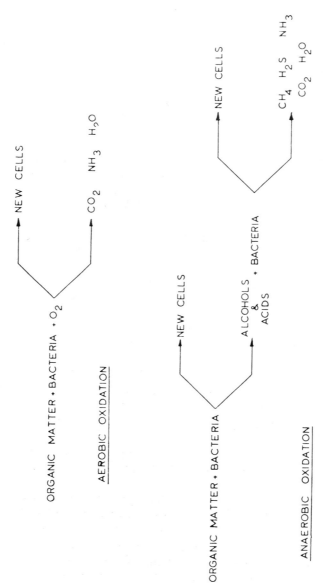

Fig. 6.1. Modes of biological oxidation.

2. Nitrogenous compounds (CHONS) containing carbon, hydrogen, oxygen, nitrogen and occasionally sulphur. The main compounds in this group are proteins which are very complex molecules, amino acids—the building blocks of proteins, and urea. The nitrogen in these compounds is liberated as ammonia on oxidation.
3. Lipids or fats (CHO) containing carbon, hydrogen and a little oxygen are only slightly soluble in water but soluble in organic solvents.

Because of the numerous organic compounds found in wastewaters it is rarely profitable or even feasible to isolate them. It is normally sufficient to determine the total amount of organic matter present (CHONS).

6.2. Enzyme Reactions

The speeds of biological reactions are controlled by enzymes, organic catalysts, which are produced by living organisms and which are capable of increasing the rate of reaction without being consumed in the process and without altering the equilibrium of the reaction.

Enzymes are highly specific, catalysing only a particular reaction and they are sensitive to environmental factors such as temperature, pH, metallic ions, etc. There are many different types of enzyme and they are classified by reference to the type of reaction which they will catalyse. Important enzyme-catalysed reactions in biochemistry are:

1. Oxidation—the addition of oxygen or the removal of hydrogen.
2. Reduction—the addition of hydrogen or the removal of oxygen.
3. Hydrolysis—the addition of water to large molecules resulting in their breakdown into smaller molecules.
4. Deamination—the removal of an NH_2 group from an amino acid or amine.
5. Decarboxylation—the removal of carbon dioxide.

6.3. Nature of Biological Growth

In the simple case of a culture of micro-organisms fed on a single occasion, the change in numbers of organisms with time follows a definite pattern as shown in Fig. 6.2. In the initial stages of growth excess food and nutrients are present so that reproduction is unrestricted and numbers

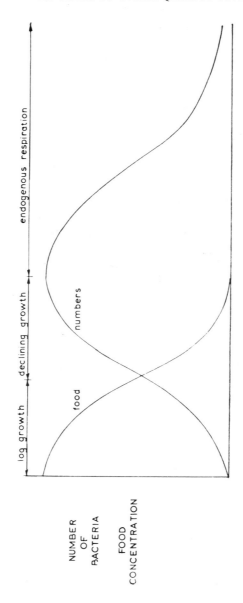

FIG. 6.2. The biological growth curve.

increase at an exponential rate. At some point the concentration of food or a particular nutrient becomes limiting and growth of the micro-organisms is increasingly restricted. In the final stage all the food has been consumed so that growth ceases and the numbers of micro-organisms decrease. Organic matter from the dead cells is utilized by the remaining organisms in auto-oxidation or endogenous respiration. This form of curve is basically true for any biological population.

For successful biological growth certain nutrients are required.

1. *Sources of carbon and nitrogen.* An empirical formula for protoplasm is $C_5H_7NO_2$ and clearly the composition of the cell will control the requirements of the organism. Thus carbon, nitrogen and to a lesser extent phosphorus are essential elements for growth. (1 kg N/15–30 kg BOD, 1 kg P/80–150 kg BOD.)

2. *Energy sources.* Micro-organisms require energy for their metabolic activities and this energy is obtained by releasing the energy of formation bound up in chemical compounds when they were originally formed from their basic constituents.

3. *Inorganic ions.* Many inorganic ions, mainly metals such as calcium, magnesium, potassium, iron, manganese, cobalt, etc., are essential to growth although only required in minute amounts. Such ions would normally be found in ample amounts in water supplies and hence also in sewage.

4. *Growth factors.* There is evidence that materials such as vitamins are required for the optimum growth of at least some types of micro-organisms.

If any of the above materials are deficient in a biological system, growth will be restricted and the shape of the growth curve will be altered.

6.4. Oxygen Demand in Aerobic Oxidation

It is of great importance in water quality control that the amount of organic matter present in the system be known and that the quantity of oxygen required for its stabilization be determined. In the case of a simple compound like glucose it is possible to write down the equation for its complete oxidation,

$$C_6H_{12}O_6 + 6O_2 \rightarrow 6CO_2 + 6H_2O$$

i.e. each molecule of glucose requires six molecules of oxygen for complete conversion to carbon dioxide and water. In the case of the more complex compounds found in most samples, e.g. proteins, etc., the reactions become more difficult to understand. In addition to the oxygen required to stabilize carbonaceous matter, there is also a considerable oxygen demand during the nitrification of nitrogenous compounds:

$$2NH_3 + 3O_2 + \text{nitrifying bacteria} \rightarrow 2NO_2^- + 2H^+ + 2H_2O$$

$$2NO_2^- + O_2 + 2H^+ + \text{nitrifying bacteria} \rightarrow 2NO_3^- + 2H^+$$

The amount of oxygen required to completely stabilize a waste could be calculated on the basis of a complete chemical analysis of the sample, but such a determination would be difficult and time-consuming. Several methods of calculating the theoretical oxygen demand knowing various characteristics of the sample have been proposed, e.g.

Ultimate oxygen demand (UOD) mg/l
$$= 2.67 \times \text{Organic carbon mg/l} + 4.57 \text{ (Org.N + Amm.N) mg/l}$$
$$+ 1.14 \, NO_2\text{–N mg/l} \qquad (6.1)$$

The chemical oxygen demand determinations using potassium permanganate or potassium dichromate measure a proportion of the UOD (a large proportion in the case of the dichromate method). Unfortunately these methods have two basic disadvantages in that they give no indication of whether or not the substance is degradable biologically and nor do they indicate the rate at which biological oxidation would proceed and hence the rate at which oxygen would be required in a biological system. Because of these disadvantages a great deal of work on waste strength measurements is done using the biochemical oxygen demand (BOD) test developed by the Royal Commission on Sewage Disposal at the turn of the century.

The BOD test measures the oxygen consumed by bacteria whilst oxidizing organic matter under aerobic conditions. The oxidation proceeds relatively slowly and is not usually complete in the standard 5-day period of incubation. Simple organic compounds like glucose are almost completely oxidized in 5 days, but domestic sewage is only about 65% oxidized and complex organic compounds might be only 40% oxidized in this period. Exertion of BOD is normally assumed to follow a first-order reaction initially, although there is evidence that biological oxidation in practice is not necessarily of this concept. In a first-order reaction the rate of oxidation is proportional to the concentration of oxidizable organic matter remaining and once a suitable population of micro-organisms has been formed the rate of reaction is controlled only by the amount of food available, i.e.

$$\frac{dL}{dt} = -KL \qquad (6.2)$$

where L = concentration of organic matter remaining or ultimate BOD,
t = time,
K = constant.

Integrating,
$$\frac{L_t}{L} = e^{-Kt} \qquad (6.3)$$

where L_t is the BOD remaining at time t (Fig. 6.3).

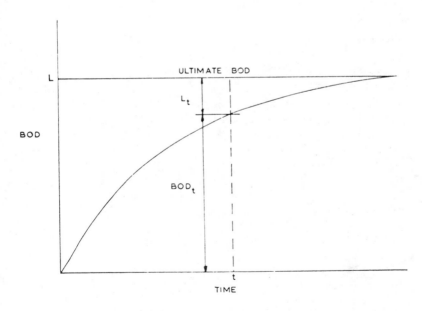

Fig. 6.3. Basis of BOD.

It is conventional to use $\log_{10}$ rather than $\log_e$ and this can be achieved by changing the constant

$$\frac{L_t}{L} = 10^{-kt} \tag{6.4}$$

where $k = 0.4343\,K$. k is termed the rate constant.

The normal concern is with oxygen taken up, i.e. BOD_t, rather than with oxygen demand remaining, thus

$$BOD_t = L - L_t = L(1 - 10^{-kt}) \tag{6.5}$$

The value of k governs the rate of oxidation, as shown in Fig. 6.4, and may be used to characterize the biological degradability of a substance. For sewage, k is about $0.17/d$ at a temperature of $20°C$. For another temperature T_L, new values can be found from

$$k_T = k_{20} \times 1.047^{(T-20)} \tag{6.6}$$

$$L_T = L_{20}[1 + 0.02(T - 20)] \tag{6.7}$$

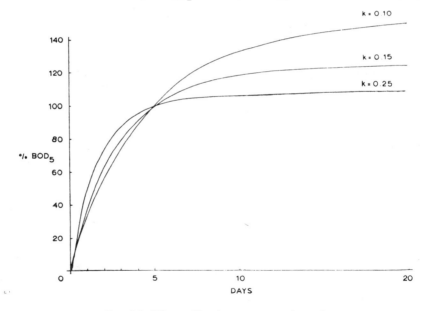

FIG. 6.4. Effects of k values on oxygen demand.

The determination of a single 5-day BOD does not of course permit calculation of L and K values for the sample. To obtain this information it is necessary to carry out determinations of BOD over periods of, say, 1 to 6 days and then subject the data to some form of curve fitting process. A simple fitting technique due to Thomas[1] relies on the fact that the expansions of

$$(1 - e^{-Kt}) \quad \text{and} \quad Kt\left(1 + \frac{Kt}{6}\right)^{-3}$$

are very similar. The expression

$$\text{BOD}_t = L(1 - e^{-Kt})$$

can thus be approximated by

$$\text{BOD}_t = LKt\left(1 + \frac{Kt}{6}\right)^{-3}$$

i.e.
$$\left(\frac{t}{\text{BOD}_t}\right)^{1/3} = (KL)^{-1/3} + \left(\frac{K^{2/3}}{6L^{1/3}}\right)t \qquad (6.8)$$

By plotting $\left(\dfrac{t}{\text{BOD}_t}\right)^{1/3}$ against t a straight line should be obtained with the intercept $(a) = (KL)^{-1/3}$ and the slope $(b) = \dfrac{K^{2/3}}{6L^{1/3}}$

Hence
$$K = \frac{6b}{a} \qquad (6.9)$$

and
$$L = \frac{1}{Ka^3} \qquad (6.10)$$

In practice the shape of the BOD curve is modified by the effect of oxygen required for nitrification (Fig. 6.5), but due to the slow rate of growth of nitrifying bacteria this effect is not normally important until 8–10 days have elapsed with raw waste samples. However, in the case of treated effluents the effect of nitrification may become apparent after a day or two due to the presence of large numbers of nitrifying bacteria in the effluent. Nitrification can be inhibited in BOD samples by the addition of allyl thiourea (ATU),

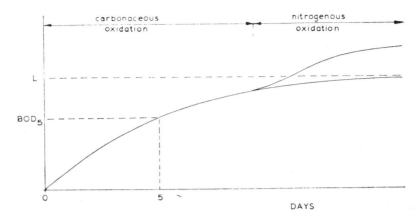

FIG. 6.5. Typical BOD curve.

so that only carbonaceous oxygen demand is measured. It should be remembered, however, that nitrification does exert a significant oxygen demand. It is sometimes argued that the presence of nitrates in an effluent provides an oxygen reservoir which may be utilized if further pollution occurs in the river. Unfortunately, however, oxygen from nitrates is only released when the DO falls below 1.0 mg/l, by which time much of the life in the water has been killed.

The BOD test was originally intended to simulate conditions which occurred following the discharge of an effluent to a river, and whilst there may be some resemblance between the conditions of the test and conditions in a surface water there is little relation between the test conditions and those which prevail in a biological treatment plant. The BOD test uses a small culture of micro-organisms to stabilize organic matter in quiescent conditions and constant temperature with a limited DO supply. In a biological treatment plant high concentrations of micro-organisms are continuously agitated to keep them in contact with the concentrated substrate and an excess of DO is supplied. A further disadvantage of the BOD determination is that the results give no indication of the rate of oxygen uptake unless BODs are determined at daily intervals over a period instead of the standard 5-day period.

Attempts to provide a more realistic approach to the measurement of oxygen demand in treatment plants have utilized the principle of the respirometer. In a simple large volume respirometer the sample is contained in a sealed flask containing an absorption boat holding potassium hydroxide solution so that during aerobic oxidation the carbon dioxide produced is absorbed and there is a resulting drop in pressure in the flask due to the consumption of oxygen. Connections to the flask permit indication of pressure drop in the flask (compared with a reference pressure) and the addition of measured volumes of oxygen to return the flask pressure to the standard conditions. Agitation of the sample is provided by a magnetic stirrer and the sample and reference flasks are immersed in a constant-temperature bath. Developments of this simple apparatus permit automatic recording of oxygen uptake over prolonged periods. Whilst the respirometer is of considerable value for research purposes, its complexity is such that the BOD test is likely to remain the standard determination for a long time.

It should be remembered that any biological test will give erroneous results in the presence of toxic materials.

6.5. Anaerobic Oxidation

With certain strong organic wastes, e.g. sludges, slaughterhouse discharges, etc., the oxygen requirement for aerobic stabilization is high and it becomes physically difficult to maintain aerobic conditions in the reaction vessel. In such circumstances anaerobic stabilization of the major part of the organic matter may be a suitable method of treatment in spite of its lower efficiency and slow rate of reaction. The basic difference between aerobic and anaerobic oxidation is that in the aerobic system oxygen is the ultimate hydrogen acceptor with a large release of energy. In the anaerobic system the ultimate hydrogen acceptor may be nitrate, sulphate or various organic compounds, resulting in a much lower release of energy. Complete stabilization of organic matter cannot be achieved anaerobically, and it is normally necessary to treat the anaerobic plant effluent further by aerobic means.

As will be seen in Fig. 6.1, anaerobic oxidation is a two-stage process and as a result has certain operational problems. The acid-forming bacteria which carry out the first stage of the breakdown are fairly adaptable as regards environmental conditions, but the methane-formers responsible for the second stage are more sensitive. In particular the methane-formers will only operate in the pH range 6.5–7.5. It is thus important to control conditions to suit the methane bacteria. Overproduction of acids by the fast-acting acid-formers can rapidly result in a low pH, thus stopping the action of the methane-formers and leaving the reaction at a point where particularly unpleasant and odoriferous compounds have been produced. Further production of acid will lower the pH to such a level that even the acid-formers are inhibited and all action will cease. Matters can then only be rectified by pH correction with chemicals, usually lime, but prevention of such mishaps is a better solution and is achieved by careful observation of pH and volatile acids concentration. Both types of bacteria prefer warm conditions, and optimum temperatures for anaerobic oxidation are about 35°C or 55°C.

Reference

1. THOMAS, H. A., Graphical determination of BOD rate constants. *Wat. Sew. Wks*, **97**, 1950, 123.

Further Reading

JENKINS, D., The use of manometric methods in the study of sewage and trade wastes. In *Waste Treatment* (Ed. Isaac, P. C. G.), Pergamon Press, Oxford, 1960, p. 99.

MCKINNEY, R. E., *Microbiology for Sanitary Engineers*, McGraw-Hill, New York, 1962.

MONTGOMERY, H. A. C., The determination of biochemical oxygen demand by respirometric methods. *Wat. Res.* **1**, 1967, 631.

SIMPSON, J. R., Some aspects of the biochemistry of aerobic organic waste treatment, *and* Some aspects of the biochemistry of anaerobic digestion. In *Waste Treatment* (Ed. Isaac, P. C. G.), Pergamon Press, Oxford, 1960, pp. 1 and 31.

TEBBUTT, T. H. Y. and BERKUN, M., Respirometric determination of BOD. *Wat. Res.* **10**, 1976, 613.

Problems

1. The analysis of a wastewater is:

Organic carbon	325 mg/l
Org.N	50 mg/l
Amm.N	75 mg/l
NO_2–N	5 mg/l

Calculate the ultimate oxygen demand.   (1445 mg/l)

2. Laboratory determinations on an industrial waste indicate that its ultimate BOD is 750 mg/l and the k value at 20°C is 0.20/d. Calculate the 5-day BOD. What would be the 5-day BOD if the k value dropped to 0.1/d? (675 mg/l, 510 mg/l)

3. Three samples all have the same 5-day BOD of 200 mg/l, but their k values are 0.10, 0.15, and 0.25/d. Determine the ultimate BOD for each sample. (295 mg/l, 244/mg/l, 212 mg/l)

4. A domestic sewage has a 5-day BOD at 20°C of 240 mg/l. If the k value is 0.1/d, determine the BOD at 1 and 5 days at 13°C. (46.5 mg/l, 171 mg/l)

5. A series of BOD (ATU) determinations was made on a sample to enable calculation of the ultimate BOD and rate constant. Incubation was carried out on a 5% dilution of the sample at 20°C when the initial saturation DO for samples and blanks was 9.10 mg/l.

Day	Final DO in sample (mg/l)	Final DO in dilution water (mg/l)
1	7.10	9.00
2	6.10	9.00
3	5.10	8.90
4	4.20	8.90
5	3.90	8.80
6	3.50	8.70
7	3.00	8.60

Use the Thomas method to calculate L and k values for the sample. (126 mg/l, 0.145/d)

CHAPTER 7

Water Pollution and its Control

It is important to appreciate that all natural waters contain a variety of contaminants arising from erosion, leaching and weathering processes. To this natural contamination is added that arising from domestic and industrial wastewaters which may be disposed of in various ways, e.g. into the sea, onto land, into underground strata or, most commonly, into surface waters.

Any body of water is capable of assimilating a certain amount of pollution without serious effects because of the dilution and self-purification factors which are present. If additional pollution occurs the nature of the receiving water will be altered and its suitability for various uses may be impaired. An understanding of the effects of pollution and of the control measures which are available is thus of considerable importance to the efficient management of water resources.

7.1. Types of Pollutant

Contaminants behave in different ways when added to water. Non-conservative materials including most organics, some inorganics and many micro-organisms are degraded by natural self-purification processes so that their concentrations reduce with time. The rate of decay of these materials is a function of the particular pollutant, the receiving water quality, temperature and other environmental factors. Many inorganic substances are not affected by natural processes so that these conservative pollutants can only have their concentrations reduced by dilution. Conservative pollutants are often unaffected by normal water and wastewater treatment processes so that their presence in a particular water source may limit its use.

As well as the classification into conservative or non-conservative characteristics the following properties of pollutants are of importance.

1. Toxic compounds which result in the inhibition or destruction of biological activity in the water. Most of these materials originate from industrial discharges and would include: heavy metals from metal finishing and plating operations, moth repellents from textile manufacture, herbicides and pesticides, etc. Some species of algae can release potent toxins and cases have been recorded where cattle have died after drinking water containing algal toxins.

2. Materials which affect the oxygen balance of the water.
    (a) Substances which consume oxygen, these may be organic materials which are biochemically oxidized or inorganic reducing agents.
    (b) Substances which hinder oxygen transfer across the air–water interface. Oils and detergents can form protective films at the interface which reduce the rate of oxygen transfer and may thus amplify the effects of oxygen-consuming substances.
    (c) Thermal pollution can upset the oxygen balance because the saturation DO concentration reduces with increasing temperature.

3. Inert suspended or dissolved solids in high concentrations can cause problems, e.g. china-clay washings can blanket the bed of a stream preventing the growth of fish food and removing fish from the vicinity as effectively as does a direct poison. The discharge of saline mine-drainage water may render a river unsuitable for water-supply purposes.

7.2. Self-purification

In a natural water, self-purification exists in the form of a biological cycle (Fig. 7.1) which is able to adjust itself, within limits, to changes in the environmental conditions. In a low organic-content stream there is little nutrient material to support life so that although many types of organisms will be present there are only relatively low numbers of each type. In streams with high organic content it is likely that the DO level will be depressed producing conditions unsuitable for animals and higher plant life. In these circumstances bacteria will predominate although given sufficient time the

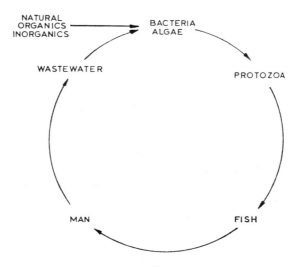

Fig. 7.1. The self-purification cycle.

organic matter will be stabilized, the oxygen demand will fall and a full range of life forms will appear again.

Self-purification involves one or more of the following processes:
1. Sedimentation, possibly assisted by biological or mechanical flocculation. The deposited solids will form benthal deposits which if organic, will decay anaerobically and which, if resuspended by flood flows, can exert sudden high oxygen demands on the system.
2. Chemical oxidation of reducing agents.
3. Bacterial decay due to the generally inhospitable environment for enteric and pathogenic bacteria in natural waters.
4. Biochemical oxidation which is normally by far the most important process. To prevent serious pollution it is important that aerobic conditions are maintained, this means that the balance between oxygen consumed by BOD and supplied by reaeration from the atmosphere is not drastically disturbed.

Reaeration

In the absence of any external mixing the concentration of a gas dissolved in water will eventually become uniform due to molecular diffusion. The rate of diffusion is proportional to the concentration gradient and is described by Fick's Law:

$$\frac{\partial M}{\partial t} = k_d A \frac{\partial C}{\partial l} \qquad (7.1)$$

where M = mass transfer in time t,
  k_d = diffusion coefficient,
  A = cross-sectional area across which transfer occurs,
  C = concentration,
  l = distance in direction of transfer.

A solution to equation (7.1) is:

$$C_t = C_s - 0.811 (C_s - C_0)(e^{-K_d} + \tfrac{1}{9}e^{-9K_d} + \tfrac{1}{25}e^{-25K_d} + \ldots) \qquad (7.2)$$

where C_0 = concentration at time 0,
  C_t = concentration at time t,
  C_s = saturation concentration,
  $K_d = \dfrac{k_d \pi^2 t}{4l^2}$.

The diffusion coefficient (k_d) is usually expressed as mm^2/s and for oxygen in water has a value of 1.86×10^{-3} mm^2/s at $20°C$.
  The solution of a gas in liquid is governed by two physical laws:

Dalton's Law of Partial Pressures

The partial pressure of a gas in a mixture of gases is the product of the proportion of that gas in the mixture and the total pressure.

Henry's Law

At constant temperature the solubility of a gas in a liquid is proportional to the partial pressure of the gas.

The rate of solution of oxygen is proportional to the saturation deficit, i.e.

$$\frac{dD}{dt} = -KD \tag{7.3}$$

hence
$$D = D_a e^{-K_2 t} \tag{7.4}$$

where D = DO deficit at time t,
   D_a = DO deficit initially,
   K_2 = reaeration constant.

Expressing the above equation in terms of DO concentrations

$$\log_e \frac{(C_s - C)}{(C_s - C_0)} = -K_2 t \tag{7.5}$$

The value of K_2 is a function of the velocity of flow, channel configuration and temperature. Rather than using the reaeration constant K_2 it is preferable to adopt a parameter which measures the reaeration rate per unit area exposed per unit DO deficit—the exchange coefficient f

$$f = K_2 \frac{V}{A} \tag{7.6}$$

where V = volume of water below interface,
   A = area of air–water interface.
$K_2(V/A)$ is termed the aeration depth and has units of velocity. Table 7.1 gives typical values of the exchange coefficient. Studies on a number of British rivers[2] have shown that the value of f at 20°C can be predicted from the formula

$$f \text{ mm/h} = 7.82 \times 10^4 \, U^{0.67} \, H^{-0.85} \tag{7.7}$$

where U = velocity of water m/s,
   H = mean depth of flow mm.
A rise in temperature of 1 degree increases the value of f by about 2% and similarly a fall in temperature decreases the rate of reaeration.

One of the problems in the study of rivers is to determine the reaeration characteristics of a stream. Solution can only occur at the air–water interface where a thin film of water is rapidly saturated and further reaeration is controlled by the diffusion of oxygen throughout the main

TABLE 7.1. TYPICAL VALUES OF THE EXCHANGE
COEFFICIENT f

Situation	f mm/h
Stagnant water	4–6
Water in channel at 0.6 m/min	10
Sluggish polluted river	20
Thames Estuary	55
Water in channel at 10 m/min	75
Open sea	130
Water in channel at 15 m/min	300
Turbulent Lakeland beck	300–2000
Water flowing down 30° slope	700–3000

After Klein[1].

body of water which is a slow process. In a turbulent stream this saturated surface layer is broken up and reaeration can proceed more rapidly (Fig. 7.2).

Field determination of the reaeration characteristics of a stream[3] involves partial deoxygenation of the stream with a reducing agent (sodium sulphite plus cobalt catalyst) and measuring DO uptake at stations

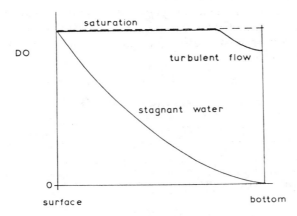

FIG. 7.2. Effect of turbulence on DO profiles.

downstream. Assuming there is no significant BOD or photosynthesis in the reach,

$$f = \frac{V}{t} \frac{1}{A} \log_e \frac{(C_s - C_1)}{(C_s - C_2)} \tag{7.8}$$

where C_1 and C_2 are DO concentrations at two stations downstream of the reagent addition with a time of flow between them of t ($V/t = Q$).

The Sag Curve

The situation which occurs in a stream receiving a single pollution load is shown in Fig. 7.3. If the stream is originally saturated with DO the BOD uptake curve for the mixture of effluent and stream water gives the cumulative deoxygenation of the stream. As soon as BOD begins to be exerted the DO falls below saturation and reaeration starts. With increasing saturation deficit the rate of reaeration increases until a critical point is reached where the rates of deoxygenation and reaeration are equal. At the critical point, minimum DO is reached and as further time passes the DO will increase. Assuming that the only processes involved are BOD removal by biological oxidation and DO replenishment by reaeration from the atmosphere the Streeter–Phelps equation was derived

$$\frac{dD}{dt} = K_1 L - K_2 D \tag{7.9}$$

where D = DO deficit at time t,

L = ultimate BOD,

K_1 = BOD reaction rate constant,

K_2 = reaeration constant.

Integrating and changing to base 10 ($k = 0.4343\,K$)

$$D_t = \frac{k_1 L_a}{k_2 - k_1} (10^{-k_1 t} - 10^{-k_2 t}) + D_a 10^{-k_2 t} \tag{7.10}$$

D_a and L_a are values at $t = 0$ ($L_t = L_a 10^{-k_1 t}$).

The critical point, i.e. the point of maximum deficit is given by

$$\frac{dD}{dt} = 0 = K_1 L - K_2 D_c \tag{7.11}$$

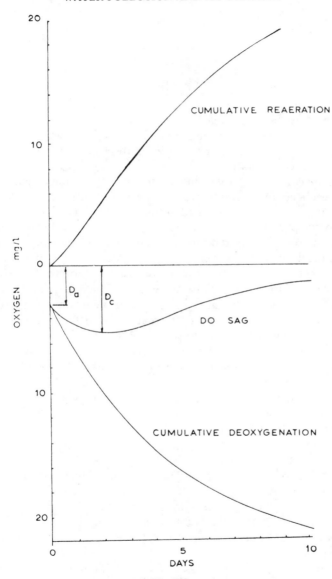

Fig. 7.3. The DO sag curve.

Hence
$$D_c = \frac{k_1}{k_2} L_a 10^{-k_1 t_c}$$
(7.12)

$$t_c = \frac{1}{k_2 - k_1} \log \left\{ \frac{k_2}{k_1} \left[1 - \frac{D_a(k_2 - k_1)}{L_a k_1} \right] \right\}$$
(7.13)

where D_c = critical deficit reached at time t_c.

This equation may be used in a number of ways although in theory it is only valid when there is no change in dilution or pollution load in the stretch under consideration.

For more complex river systems, a step-wise calculation can be adopted, treating each section of river between changes in pollution and/or flow as an individual problem. The calculated output BOD and DO from this section then provide part of the input data for the next section downstream. The calculation process can be repeated as often as required to make a predictive mathematical model for a particular river system. Such a model can also include other non-conservative pollutants using theoretical or empirical decay relationships analogous to the Streeter–Phelps expression for DO.

Other factors may come into play in the oxygen balance, probably the most important being deposition of organic matter from suspension and possible later resuspension due to scour of bottom mud. The contribution of bottom muds to oxygen demand can be considerable in shallow wide channels but relatively small in narrow deep channels. Flood flows can resuspend muds giving a very high oxygen demand.

Other factors which may influence the DO sag include:
1. BOD addition in surface runoff.
2. Removal of DO by diffusion into bottom mud to satisfy oxygen demand.
3. BOD addition by diffusion of soluble organics from bottom deposits.
4. Removal of DO by purging action of gases released from bottom deposits.
5. Addition of DO by photosynthetic activities of plants.
6. Removal of DO by plants during night.
7. Continuous redistribution of DO and BOD by longitudinal dispersion.

Dobbins[4] has produced a method of allowing in part at least for the above factors. It was found that longitudinal dispersion of BOD and DO in most freshwater streams was negligible. Accurate measurement of surface reaeration was most important.

7.3. Toxic Materials

Fish are usually employed as sensitive indicators of toxic pollution but the situation is complicated because various environmental factors can considerably affect the toxicity of a particular material. As far as fish are concerned the two most important environmental factors are DO and temperature. Fish require a certain minimum oxygen supply for normal activity ranging from about 1.5 mg/l for certain coarse fish to 5 mg/l for game fish. At or near these limiting DO levels the activity of fish may be impaired so that their sensitivity to poisonous materials is often increased. Certain poisons such as the heavy metals interfere with respiration so that their harmful properties are enhanced at low DO. Thus a reduction in DO to 50% saturation will reduce the concentration at which a heavy metal is toxic to about 70% of the concentration which is toxic in oxygen-saturated water.

The metabolic rate of fish is closely linked with temperature so that a rise of 10 deg will increase the oxygen requirement by 2 or 3 times. Unfortunately the saturation concentration of DO falls with increasing temperature so that the effect of raising the temperature is to raise the oxygen requirement whilst simultaneously reducing the available oxygen supply. As a rough guide it can be taken that a 10 deg rise in temperature will approximately halve the concentration at which material is toxic.

Another factor which can have a considerable effect on toxicity is pH. A good example of this being found in the behaviour of ammonium compounds which are relatively innocuous at low pH values. Under alkaline conditions, however, ammonia can be quite harmful to fish, a rise in pH from 7.4 to 8.0 can halve the toxic concentration. It appears that unionized ammonia is the toxic form, ionized ammonia, which predominates at low pH values, being much less toxic. There is evidence to show that unionized substances are more readily absorbed by fish than the ionized forms.

The presence of dissolved salts in water is a further factor which can influence the toxicity of certain substances. The presence of calcium ions in solution will considerably reduce the toxic effect of heavy metals such as lead and zinc. High concentrations of sodium, calcium and magnesium prevent the toxic effects of heavy metals probably by forming complexes with the heavy metals. For example, 1 mg/l of lead in a soft water may be rapidly fatal to fish, but in a hard water of, say, 150 mg/l calcium hardness 1 mg/l of lead will not be harmful.

The effect of potentially toxic materials in rivers is normally measured by their action on fish as demonstrated by some form of bioassay. The procedure involves the use of a series of dilutions of the suspect material to

TABLE 7.2. SOME COMPOUNDS TOXIC TO FISH

Material	Occurrence	Approx. LD_{50} mg/l
Acridine	Coal-tar wastes	0.7–1.0
Aldrin	Insecticide	0.02
Alkyl benzene sulphonate	Sewage effluent	3–12
Ammonia	Sewage effluent	2–3
	Gas works effluent	
Chloramine	Chlorinated effluents	0.06
Chlorine	Chlorinated effluents	0.05–0.2
Copper sulphate	Metal processing	0.1–2.0
	Algal control of reservoirs	
Cyanide	Gas liquor	0.04–0.1
	Plating wastes	
DDT	Insecticide	< 0.1
Detergents synthetic (packaged)	Sewage effluent	15–80
Fluoride	Aluminium smelting	2.5–60
Gammexane	Insecticide	0.035
Hydrogen sulphide	Bottom muds, sludge	0.5–1.0
Methyl mercaptan	Oil refineries	1.0
	Wood pulp processing	
Naphthalene	Coal-tar wastes	10–20
	Gas liquor	
Parathion	Insecticide	0.2
Potassium dichromate	Flow gauging	50–500
Silver nitrate	Photographic wastes	0.004
Zinc	Galvanizing	1–2
	Rayon manufacture	

Note: These figures are intended only as a guide, the actual LD_{50} in any particular situation will depend on environmental factors, the species of fish involved and the duration of the exposure.

which test fish are exposed under standard conditions. The prescribed measure of acute toxicity is the median tolerance limit (TL_m) sometimes referred to as the 50% lethal dose (LD_{50}). This is the concentration of material under test at which 50% of the test fish are able to survive for a specified period of exposure (usually 48 or 96 h).

Various bioassay procedures are in use and because of the many variations in procedure and in environmental conditions it is not possible to state that fish can only tolerate a certain concentration of a particular material. However, Table 7.2 gives some typical values of toxic levels as a guide.

When considering raw waters for potable supply the presence of toxic substances must always be seen as a potential hazard. The use of lowland rivers for potable supply inevitably implies increased potential danger because of accidental discharges of toxic materials. Unfortunately it is not feasible to analyse a raw water for all known toxic compounds and reliance must largely be placed on rapid reporting of accidental discharges of such compounds. The recommended 7-day bankside storage of raw water from lowland rivers provides a degree of safeguard, but there is nevertheless a need for some form of monitoring device to warn of the presence of toxic material. A universal monitoring device is not likely to be available, but a number of techniques using fish or micro-organisms offer at least some degree of warning of acute levels of toxic material. The question of trace levels of toxic compounds is much more difficult of solution.

7.4. Overall Effects of Pollution

When considering pollution by wastewaters there are of course other effects than the creation of DO deficits. Depending on the dilution available there will be significant increases in dissolved solids, organic content, nutrients such as N and P, colour and turbidity. All of these constituents may give rise to undesirable changes in water quality particularly as regards downstream abstraction. Nutrient build-up is a serious problem in lakes and very slow-moving waters but is not likely to be so troublesome with rivers. It should be remembered, however, that in many river abstraction schemes for water supply, raw water is stored in large shallow reservoirs prior to treatment. Even quite low nutrient contents in the water can result in prolific algal growths giving a water much more difficult to treat than the original river water.

All lakes undergo a natural change in their characteristics which in the absence of man may take thousands of years. A lake in a 'wilderness' catchment receives inflow from largely barren surroundings and thus collects little in the way of organic food and inorganic nutrients. Such nutrient-deficient waters are termed oligotrophic and are characterized by low TDS levels, very low turbidities and small biological populations. As the catchment becomes older (in almost geological time scale) there is a gradual increase in nutrient levels and hence in biological productivity, with a consequent deterioration in water quality. Eventually, as nutrient levels and biological production increase, the water becomes nutrient-rich or eutrophic. Nutrients are recycled and in extreme cases the water may become heavily polluted by vegetation, low DO levels will occur due to rotting plants and during darkness anaerobic conditions may well exist. The eventual fate of all lakes is to become eutrophic, but the rate at which this end point is reached can be greatly accelerated by artificial enrichment due to human activities. Nitrogen and phosphorus are the most important nutrients in the context of eutrophication and since some algae can fix atmospheric nitrogen it is generally accepted that phosphorus is the limiting nutrient in water. The level of phosphorus above which algal growth becomes excessive depends upon many factors, but in UK conditions waters with a winter phosphate level less than $5 \mu g/l$ are unlikely to exhibit eutrophic tendencies. Phosphates occur in sewage effluents due partly to human excretion and partly to their use in synthetic detergents.

Serious pollution which often occurs in industrialized areas can have very profound effects on a river system and reduction of river pollution in such a system is inevitably an expensive operation usually taking many years to achieve. Ideally it would be desirable for every river to be unpolluted, full of fish and aesthetically pleasing. In an industrialized country it became economically impossible to prevent all river pollution and it is necessary to take an overall view of water resources and to classify rivers as suitable for particular purposes.

River pollution is clearly undesirable for many reasons:
1. Contamination of water supplies—additional load on treatment plants.
2. Restriction of recreational use.

3. Effect on fish life.
4. Creation of nuisances—appearance and odour.
5. Hindrance to navigation by banks of deposited solids.

A typical water use classification might thus be (in decreasing order of quality requirements):

1. Domestic water supply.
2. Industrial water supply.
3. Commercial fishing.
4. Irrigation.
5. Recreation and amenity.
6. Transportation.
7. Waste disposal.

Each use has specific requirements for quality and quantity of water and some uses may be incompatible. Irrigation is a consumptive use in that water used in this way does not find it way back into the river system. Considerable volumes of cooling water are lost by evaporation. The others are not in general consumptive although they usually have a detrimental effect on quality. Thus water abstracted for domestic supply is returned as sewage effluent. The conservation of water resources depends on multi-purpose use of water wherever possible.

7.5. Groundwater Pollution

Although groundwaters are normally effectively purified, as far as suspended matter is concerned, by the straining action of the rock as water percolates through it, soluble impurities are not so readily removed, although there is the possibility of ion-exchange reactions.

Nitrogen compounds in agricultural drainage and in effluents discharged to soakaways are responsible for high nitrate levels in some groundwaters. The use of soakaways for the disposal of domestic and industrial effluents may impair groundwater quality unless there is an impermeable stratum between the disposal area and the aquifer. Similar considerations apply to the siting of refuse tips for both domestic and industrial solid wastes. These potential hazards are much increased if the strata is fissured, since polluted water may then rapidly reach the aquifer.

Organic matter entering groundwater reservoirs will only be stabilized slowly because the oxygen demand rapidly deoxygenates the water and there is no source of replacement oxygen. Anaerobic conditions then ensue and constituents of the aquifer, e.g. iron, may then readily dissolve in the water, causing further quality problems.

It is also important to note that although rocks may strain out suspended matter, this process is likely to have a deleterious effect on the permeability of the strata. Such considerations are clearly of considerable importance in the case of recharge operations.

7.6. Pollution of Tidal Waters

For communities with access to tidal waters the possibility of their utilization as a wastewater disposal facility is clearly attractive. The potential for dilution and dispersion of pollutants in the open sea is considerable and a number of self-purification reactions operate in the marine environment. In many parts of the world it has therefore become common to consider tidal waters as an infinite sink for the disposal of unwanted materials. Unfortunately, tidal waters can have widely varying characteristics so that the assessment of the effects of pollution requires great care. At one extreme, the upper reaches of a tidal estuary may have similar pollution-assimilating characteristics to those of the non-tidal reach of the river. In some cases narrow tidal estuaries exhibit an oscillatory movement so that a discharge may take several days to travel a relatively short distance to the open sea. The other extreme could be considered as the situation where the discharge is made to deep open water with a strong tidal current. Although environmental groups may object to the discharge of wastes to tidal waters there is little evidence to support objections to properly designed outfalls. To ensure satisfactory disposal it is essential to ensure that sufficient dilution is available to reduce the concentration of wastewater at the surface to very low levels. This may involve the construction of an outfall sewer several kilometres out to sea with a series of diffusers at the seaward end to aid initial dispersion. Particular care is necessary when discharging wastewaters in the vicinity of bathing beaches and shell-fish beds and in these circumstances it may be justifiable to install full sewage-treatment facilities, similar to those employed for discharges to non-tidal waters.

7.7. Control of Pollution

Because of the need to reconcile the various demands on water resources most countries have pollution-control bodies to maintain and hopefully improve water quality. In this context it is useful to quote the EEC definition that water pollution means "the discharge by man of substances into the aquatic environment the results of which are such as to cause hazards to human health, harm to living resources and aquatic ecosystems, damage to amenities or interfere with other legitimate uses of water". It follows that for a discharge to be termed polluting there must be evidence of actual harm or damage.

In England and Wales all aspects of the water cycle come under the control of regional water authorities which are responsible for the whole catchment of a major river or rivers. For pollution control these authorities have powers laid down by the Control of Pollution Act, 1974 and earlier legislation. These enable them to grant consents for the discharge of effluents to surfacewaters and groundwaters subject to standards imposed by the authority as to composition, e.g. pH, BOD, SS, Amm.N, etc., and rate of flow.

When establishing methods for the control of water pollution, standards can be based either on the quality required in the receiving water (the River Quality Objective approach) or they can be applied directly to the effluent without reference to the receiving water (the Emission Standard approach). The Quality Objective method appears logical but can cause problems when a new discharge is added to the system since either all existing discharge levels must be revised downward or the new discharge may be faced with the attainment of an impossibly high standard. There could be inequalities in the degree of treatment required for similar wastewaters discharged to different sections of the same river. A downstream effluent could require more treatment because the dilution water would be of a lower quality as a result of the upstream discharge. The Emission Standard concept is administratively convenient in that the standard is applied to all similar discharges but it has the disadvantage that no allowance is made for the self-purification characteristics of the receiving water nor of its downstream use. The compromise of adopting emission standards based on the use of the receiving water has the merit of often being easier to enforce than receiving water standards but does not of itself ensure the maintenance

TABLE 7.3. UK NWC RIVER QUALITY CLASSIFICATION

River class	Quality criteria	Remarks	Current potential uses
1A	(i) DO saturation greater than 80%. (ii) BOD not greater than 3 mg/l. (iii) Ammonia not greater than 0.4 mg/l. (iv) Where the water is abstracted for drinking water, it complies with requirements for A2* water. (v) Non-toxic to fish in EIFAC terms (or best estimates if EIFAC figures not available).	(i) Average BOD probably not greater than 1.5 mg/l. (ii) Visible evidence of pollution should be absent.	(i) Water of high quality suitable for potable supply abstractions and for all other abstractions. (ii) Game or other high class fisheries. (iii) High amenity value.
1B	(i) DO greater than 60% saturation. (ii) BOD not greater than 5 mg/l. (iii) Ammonia not greater than 0.9 mg/l. (iv) Where water is abstracted for drinking water, it complies with the requirements for A2* water. (v) Non-toxic to fish in EIFAC terms (or best estimates if EIFAC figures not available).	(i) Average BOD probably not greater than 2 mg/l. (ii) Average ammonia probably not greater than 0.5 mg/l. (iii) Visible evidence of pollution should be absent. (iv) Waters of high quality which cannot be placed in Class 1A because of high proportion of high quality effluent present or because of the effect of physical factors such as canalization, low gradient or eutrophication.	Water of less high quality than Class 1A but usable for substantially the same purposes.
2	(i) DO greater than 40% saturation. (ii) BOD not greater than 9 mg/l. (iii) Where water is abstracted for drinking water, it complies with requirements for A3* water.	(i) Average BOD probably not greater than 5 mg/l. (ii) Water not showing physical signs of pollution other than humic coloration and a little foaming below weirs.	(i) Waters suitable for potable supply after advanced treatment. (ii) Supporting reasonably good coarse fisheries. (iii) Moderate amenity value.

3

(iv) Non-toxic to fish in EIFAC terms (or best estimate if EIFAC figures not available).

(i) DO greater than 10% saturation.
(ii) Not likely to be anaerobic.
(iii) BOD not greater than 17 mg/l.**

Waters which are polluted to an extent that fish are absent or only sporadically present. May be used for low-grade industrial abstraction purposes. Considerable potential for further use if cleaned up.

4

(i) DO less than 10% saturation.
(ii) Likely to be anaerobic at times.

Waters which are grossly polluted and are likely to cause nuisance.

X

DO greater than 10% saturation.

Insignificant watercourses and ditches not usable, where objective is simply to prevent nuisance developing.

Notes: (a) Under extreme weather conditions (e.g. flood, drought, freeze-up), or when dominated by plant growth, or by aquatic plant decay, rivers usually in Classes 1, 2 and 3 may have BODs and dissolved oxygen levels, or ammonia content outside the stated levels for those Classes. When this occurs the cause should be stated along with analytical results.

(b) The BOD determinations refer to 5-day carbonaceous BOD (ATU). Ammonia figures are expressed as NH_4.

(c) In most instances the chemical classification given above will be suitable. However, the basis of the classification is restricted to a finite number of chemical determinands and there may be a few cases where the presence of a chemical substance other than those used in the classification markedly reduces the quality of the water. In such cases, the quality classification of the water should be downgraded on the basis of the biota actually present, and the reasons stated.

(d) EIFAC (European Inland Fisheries Advisory Commission) limits should be expressed as 95% percentile limits.

* EEC category A2 and A3 requirements are those specified in the EEC Council Directive of 16 June 1975 concerning the Quality of Surface Water intended for Abstraction of Drinking Water in the Member States.

** This may not apply if there is a high degree of reaeration.

of water quality under changing effluent discharge conditions.

Water-pollution control in the UK has been largely based on the pioneering work of the Royal Commission on Sewage Disposal which in its Eighth Report (1912) proposed the adoption of effluent standards related to the quality and volume of the dilution water. From its studies the Commission suggested that a BOD of 4 mg/l in a watercourse was a limit which if exceeded would indicate a significant degree of pollution. The recommendations on effluent standards with a 75 percentile norm of 20 mg/l BOD and 30 mg/l SS were allied with the need to provide sufficient dilution to prevent the downstream BOD exceeding 4 mg/l. Thus a clean river with BOD 2 mg/l would require 8 times the flow of a 20 mg/l BOD effluent to ensure that the downstream BOD did not exceed 4 mg/l. Unfortunately, in many cases where the recommendations were implemented, insufficient dilution was available so that the BOD limit was exceeded. However, with a non-conservative parameter like BOD the self-purification characteristics of the receiving water have a major effect on the results of a discharge and in practice many waters can tolerate 4 mg/l BOD without serious consequences.

The objectives of regional water authorities as stated by Lester[5] are:
  (i) The provision of a sufficient quantity of water of suitable quality for domestic and industrial abstractions.
 (ii) The safeguarding of public health.
(iii) The maintenance and improvement of fisheries.
(iv) The maintenance and restoration of water quality and conservation of flora and fauna associated with water.

To achieve these objectives a policy of local emission standards allied to a water quality classification scheme developed by the National Water Council (Table 7.3) has been adopted in England and Wales. The eventual aims are to ensure no deterioration in water quality, to eliminate Class 4 waters and to upgrade Class 3 waters to Class 2. The effluent standards necessary to satisfy the objectives are usually based on BOD, SS, and Amm.N concentrations for sewage discharges, using 95 percentile values. A similar philosophy is applied to industrial wastewater discharges with local effluent standards being set on the basis of the predicted effect of the particular contaminants on the receiving water. For many industrial wastewaters, treatment in admixture with domestic sewage is often an attractive proposition and the cost of the treatment is recovered by means of a charge determined from the following type of expression:

$$\text{Charge per m}^3 \text{ of effluent} = R + V + \frac{O_i}{O_s}B + \frac{S_i}{S_s}S \qquad (7.14)$$

where R = reception and conveyance charge/m^3,
   V = volumetric and primary treatment cost/m^3,
   O_i = COD (mg/l) of industrial effluent after settlement,
   O_s = COD (mg/l) of settled sewage,
   B = biological oxidation cost/m^3 settled sewage,
   S_i = total SS (mg/l) of industrial effluent,
   S_s = total SS (mg/l) of crude sewage,
   S = treatment and disposal costs of primary sludge/m^3.
Typical UK values for the above parameters (1982 prices) are $R = 3.3$ p/m^3, $V = 2.6$ p/m^3, $B = 4.6$ p/m^3, $S = 2.5$ p/m^3, $O_s = 370$ mg/l, $S_s = 340$ mg/l. This type of charging scheme encourages the industrial dischargers to take steps to reduce the volume and strength of the wastewater by careful process control and, possibly, modification of processes. If industrial wastes are taken into the main drainage system it is important to ensure that the wastewater does not contain material harmful to the fabric of the sewers, to sewer workers or to the sewage-treatment processes. In some cases, therefore, it may be necessary for pretreatment to be undertaken before discharge is made to the sewer. The "polluter must pay" policy sometimes advocated for dealing with industrial waste discharges may not be altogether satisfactory unless the charges are rationally based. In some situations an industrialist may prefer to pay the cost of causing pollution as an operating expense rather than having capital invested in a treatment plant. Such an approach would be likely to have generally detrimental effects on water quality.

Similar considerations to those outlined above apply to control of groundwater pollution although here, because of the difficulty of rectifying the damage caused by pollution of an aquifer, larger factors of safety than with surfacewater discharges are often employed. Particular care must be taken to protect important aquifers and in some cases underground disposal of liquid wastes and solid waste tips with leachate problems may only be permitted if the aquifer is known to be completely isolated from the potential source of pollution.

In the case of tidal waters, discharges may be regulated on the basis of the normal physical and chemical parameters used for inland discharges suitably adjusted to allow for the available dilution. Thus in situations with adequate dilution the discharge of screened or comminuted sewage may be acceptable. When the major concern is in relation to bathing or shellfish beds the bacteriological effects of sewage pollution are likely to be the most significant. Although the health implications of bacteriological contamination of tidal waters are difficult to quantify, various authorities have produced bathing water standards based on coliform counts ranging from 100/100 ml in California to 10 000/100 ml in the EEC. Here again it would seem that local standards suited to the particular climatic and environmental conditions are more likely to be appropriate than universal standards.

It is important to appreciate that in addition to discharges from effluent outfalls, etc., there is considerable pollution from non-point sources which are often difficult or impossible to control. These non-point sources of environmental contamination are essentially surface runoff discharges from urban areas, where the pollutants include oil and rubber compounds from road surfaces and from rural areas where the main contaminants are likely to be inorganic nutrients. Uncontrolled storm-water overflows on combined sewers are also sometimes included in the non-point pollution category. When efficient control of point sources is practiced, non-point sources can contribute significant amounts of pollution and it is vital when preparing pollution-control policies to allow for this contribution since otherwise the benefits to the environment of the control policy may be overestimated.

References

1. KLEIN, L., *River Pollution. 2. Causes and Effects*, Butterworths, London, 1962, p. 237.
2. OWENS, S. M., EDWARDS, R. W. and GIBBS, J. W., Some reaeration studies in streams, *Int. J. Air. Wat. Pollut.* **8**, 1964, 469.
3. GAMESON, A. L. H., TRUESDALE, G. A. and DOWNING, A. L., Re-aeration studies in a lakeland beck. *J. Instn Wat. Engrs*, **9**, 1955, 571.
4. DOBBINS, W. E., BOD and oxygen relationships in streams. *Proc. Amer. Soc. Civ. Engrs*, **90**, 1964, SA3, 53.
5. LESTER, W. F., River quality objectives. *J. Instn Wat. Engrs Scits*, **33**, 1979, 429.

Further Reading

BISWAS, A. K. (Ed.), *Models for Water Quality Management*, McGraw-Hill, New York, 1981.

CHARLTON, J. A., The design of sea outfalls with reference to EEC amenity water pollution criteria. *J. Instn Wat. Engrs Scits*, **34**, 1980, 33.

COLE, J. A., *Groundwater Pollution in Europe*, Water Information Centre, New York, 1974.

GARNETT, P. H., Thoughts on the need to control discharges to estuarial and coastal waters. *Wat. Pollut. Control*, **80**, 1981, 172.

GAMESON, A. L. H., EEC Directive on quality of bathing water. *Wat. Pollut. Control*, **78**, 1979, 206.

LESTER, W. F., Implementation of the Control of Pollution Act, 1974. *Wat. Pollut. Control*, **79**, 1980, 165.

LUCAS, J. L. and SIMM, R. I. C., Groundwater quality in the Severn–Trent region. *Wat. Pollut. Control*, **80**, 1981, 317.

MCINTOSH, P. T. and WILCOX, J., Water pollution charging systems in the EEC. *Wat. Pollut. Control*, **78**, 1979, 183.

NATIONAL WATER COUNCIL, *Review of Discharge Consent Conditions*, NWC, London, 1977.

NATIONAL WATER COUNCIL, *River Quality—The 1980 Survey and Future Outlook*, NWC, London, 1981.

NEMEROW, N. L., *Scientific Stream Pollution Analysis*, McGraw-Hill, New York, 1974.

NORTON, M. G., The control and monitoring of sewage sludge dumping at sea. *Wat. Pollut. Control*, **77**, 178, 402.

NOVOTNY, V. and CHESTERS, G., *Handbook of Nonpoint Pollution*, Van Nostrand Reinhold, New York, 1981.

SELBY, K. H. and SKINNER, A. C., Aquifer protection in the Severn–Trent region: Policy and practice. *Wat. Pollut. Control*, **74**, 1975, 526.

SNOOK, W. G. G., Submarine pipeline and diffuser design. *Pub. Hlth Engr*, **8**, 1980, 183.

TEBBUTT, T. H. Y., A rational approach to water quality control. *Wat. Supply Managmt*, **3**, 1979, 41.

VELZ, C. J., *Applied Stream Sanitation*, Wiley–Interscience, New York, 1970.

Problems

1. A quiescent body of water has a depth of 300 mm and its DO concentration is 3 mg/l. Determine the DO concentration at the bottom after a period of 12 days if the surface is exposed to the atmosphere at a temperature of 20°C. k_d at 20°C is 1.86×10^{-3} mm^2/s. (4 mg/l)

2. A town of 20 000 people is to discharge treated domestic sewage to a stream with a minimum flow of 0.127 m^3/s and BOD 2 mg/l. The sewage d.w.f. is 135 l/person day and the *per capita* BOD contribution is 0.068 kg/d. If the BOD in the stream below the discharge is not to exceed 4 mg/l, determine the maximum permissible effluent BOD and the percentage purification required in the treatment plant. (12 mg/l, 97.5 %)

3. A stream with BOD 2 mg/l and saturated with DO has a normal flow of 2.26 m^3/s and receives a sewage effluent, also saturated with DO, of 0.755 m^3/s with BOD 30 mg/l.

Determine the DO deficits over the next 5 days and hence plot the sag curve. Calculate the critical DO deficit and the time at which it occurs. Assume temperature is 20°C throughout. Saturation DO at 20° C 9.17 mg/l, k_1 for effluent/water mixture 0.17/d, k_2 for stream 0.40/d.   (2.38 mg/l, 1.61 d)

4. A stream with a flow of 0.75 m³/s and BOD 3.3 mg/l is saturated with DO (9.17 mg/l at 20°C). It receives an effluent discharge of 0.25 m³/s, BOD 20 mg/l and DO 5.0 mg/l. Determine the DO deficit at a point 35 km downstream if the average velocity of flow is 0.2 m/s. Assume temperature is 20°C throughout. k_1 for effluent/water mixture 0.10/d, k_2 for stream 0.40/d.   (1.87 mg/l)

5. A stream with flow 4 m³/s, BOD 1 mg/l and saturated with oxygen receives at A a sewage effluent discharge of 2 m³/s with BOD 20 mg/l and DO 4 mg/l. At point B, 20 km downstream of A, a tributary with flow 2 m³/s, BOD 1 mg/l and DO 8 mg/l joins the main stream. A further distance of 20 km downstream at C the stream receives another effluent of 2 m³/s with BOD 15 mg/l and DO 6 mg/l. Determine the DO deficit at point D, 20 km downstream of C assuming constant temperature of 20°C for which the saturation DO is 9.1 mg/l. For all reaches of the stream, $k_1 = 0.1/d$ and $k_2 = 0.35/d$, velocity of flow $= 0.3$ m/s.   (2.2 mg/l)

CHAPTER 8

Quantities of Water and Wastewater

A PERSON'S basic physiological requirement for water is about 2.5 l/day although work load and climatic conditions can greatly increase this figure, largely because of the need to replace water lost by perspiration. As the standard of living increases so does the need for water and this brings as a consequence the production of wastewaters for which suitable treatment and disposal methods must be provided.

8.1. Water Demand

In addition to the water required for survival, other domestic uses of water are highly desirable, e.g. for personal hygiene, washing of utensils and clothes, etc. The amount of water used for these other domestic purposes will be governed by the availability of water in the community judged on both amount and cost bases. In very primitive communities water demands of around 2.5 l/person day have been recorded but as life styles develop, a water demand of about 10 l/person day is normal in the absence of a piped supply and where water has to be carried some distance to the house. The provision of a central stand-pipe supply in a village will probably increase the water demand to about 25 l/person day and the demand with an individual house tap in a low-income community is likely to be about 50 l/person day. In developed countries and the high-value urban housing areas of developing countries the provision of multiple taps, flush toilets, washing machines and dishwashers will greatly increase water demand for domestic purposes so that demands of several hundred litres per person day are common. Table 8.1 gives an example of current domestic water demand in the UK which suggests that about 100 l/person day would be a reasonable figure for usage in modern housing in a temperate country. The

87

TABLE 8.1. TYPICAL DOMESTIC WATER USAGE IN THE UK
(after Thackray[1])

Use	Consumption l/person day
Toilet flushing	32
Drinking, cooking and dishwashing	33
Baths and showers	17
Clothes washing	12
Garden watering	1
Car washing	1
Total	96

water closet toilet accounts for a considerable proportion of domestic use and in the UK a 9-l flush is normal although new installations will usually have a dual-flush facility, delivering about 4 l for use after urination. In other parts of the world much larger flushes are customary and these can have a marked influence on water consumption. In the USA, for example, large toilet flushes and air conditioning and garden watering requirements may produce domestic water consumptions as high as 500 l/person day.

Industrial processes consume considerable amounts of water and in manufacturing areas industrial water demand may equal or even exceed the domestic demand. Table 8.2 shows some typical industrial water demands which are, however, very much dependent upon such factors as, the age of the plant, the cost of water and the incentive for in-plant recycling. Many industrial uses of water do not require a potable supply and there is increasing use of lower-grade sources, such a sewage effluent, to satisfy at least part of the industrial demand. Industrial water demand is closely related to industrial productivity and is therefore likely to be subject to change in differing economic circumstances. Increasing costs of water and charges for wastewater collection and treatment exert considerable pressures for reductions in the water used by industry.

In many developed countries all water consumers are metered so that, at least in theory, demand could be regulated by pricing policy. However, the UK does not have a general policy of metering individual domestic supplies although the 1973 Water Act makes provision for the installation of meters and some authorities now offer metering as an option. Most domestic consumers in the UK pay for their water on the basis of a charge related

TABLE 8.2. EXAMPLES OF INDUSTRIAL WATER USAGE
(after Thackray[2] and others)

Product or service	Consumption	Units
Coal	250	l/tonne
Bread	1.3	l/kg
Meat products	16	l/kg
Milk bottling	3	l/l
Brewing	5	l/l
Soft drinks	7	l/l
Chemicals	5	l/kg
Steel rolling	1900	l/tonne
Iron casting	4000	l/tonne
Aluminium casting	8500	l/tonne
Automobiles	5000	l/vehicle
Electroplating	15 300	l/tonne
Carpets	34	l/m^2
Textile dyeing	80	l/kg
Concrete	390	l/m^3
Paper	54 000	l/tonne
Dairy farming	150	l/cow day
Pig farming	15	l/pig day
Poultry farming	0.3	l/bird day
Schools	75	l/person day
Hospitals	175	l/person day
Hotels	760	l/employee day
Shops	135	l/employee day
Offices	60	l/employee day

essentially to the size of the property and its value so that they have no incentive to economize in their use of water. It has always been argued that the cost of installing and reading meters would far outweigh the potential savings which might accrue. Indeed, evidence from other countries where domestic metering is practised is somewhat conflicting and it is by no means clear that the introduction of metering would produce a sustained reduction in water consumption. Some form of sliding-scale tariff for metered water to discourage excessive consumption is a possibility but it must be remembered that an ample supply of safe water is a primary requirement in the maintenance of public health. Poorer members of a community should not therefore be forced into undesirable restrictions on water usage by economic pressures. The justification for installing domestic water meters in the UK is thus essentially related to the provision of an equitable means of payment for the service in which the consumer has some degree of control over the size of the bill.

In a complex distribution system it is inevitable that there will be a certain amount of leakage and waste. In supply systems which are fully metered it is not unusual for about 25 % of the water entering the system to be unaccounted for and there is no reason to suppose that the losses would be any less in unmetered systems. This loss is made up of leakages, fire-fighting usage, unauthorized connections, etc.

In temperate zones almost all of the domestic water supply and much of the industrial supply find their way back into the sewers so that the dry weather flow (d.w.f.) of sewage could be expected to be of the same order as the flow of water supplied to the area. In warm climates a proportion of the water will be used for garden watering or otherwise lost by evaporation so that only 70–80 % may enter the sewers.

Treatment plants and their associated collection and distribution systems are expensive items of capital expenditure which are often designed to have a useful life of 30 years or more. For this reason and to enable efficient development and utilization of water resources it is necessary to be able to predict future water demands. Domestic water demand is the product of the *per capita* demand and the population. The *per capita* demand will tend towards a ceiling value related to the standard of living of the community and its environmental conditions. Although in many developed countries populations are more or less static in numbers the developing world is experiencing large increases in population. The economic prosperity of an area will have a considerable influence on industrial activity and hence industrial water consumption. As productivity increases, so will water consumption until the point where the cost of water becomes a significant item in overall costs and means of reducing consumption then become financially attractive. The prediction of future industrial demands for water is thus fraught with difficulties.

8.2. Population Growth

The classical biological growth curve was discussed in Chapter 6 and although this may be considered as a basis for predicting the growth of human populations many factors make this model of limited use for such purposes. Improved living standards and better medical care in developed countries have increased life expectancy by some 50% during the last 100 years. Variations in national or local economics can affect the birth rate as do wars. Changes in industry can completely alter the nature of growth in an area and the decline of an industry can cause a fall in population in the surrounding area.

In most parts of the world a census is taken at ten-year intervals and it is logical to use this information as an aid in population prediction although its accuracy in developing countries may be dubious. Various techniques can be employed to predict future populations from existing records. If the growth is believed to be linear the following expression can be used:

$$Y_m = \frac{t_m - t_2}{t_2 - t_1}(Y_2 - Y_1) + Y_2 \tag{8.1}$$

where Y_m = population at future year t_m,
$\quad Y_1$ and Y_2 are known populations in years t_1 and t_2.

If geometric growth is believed to be appropriate, equation (8.1) can be modified to

$$\log Y_m = \frac{t_m - t_2}{t_2 - t_1}(\log Y_2 - \log Y_1) + \log Y_2 \tag{8.2}$$

Some authorities prefer to adopt a procedure of fitting complex polynomial expressions to census data. The use of mathematical relationships may however give a spurious accuracy to predictions and in many cases the best method is to plot census data and extrapolate the line for the required design period using all the information available regarding future developments in the area.

8.3. Wastewater Flow

The volume and nature of wastewater depends upon the type and age of the sewage system. In old systems, damaged sewers and cracked joints may allow loss of sewage into the surrounding ground and conversely infiltration of groundwater may increase the flow of sewage. Older communities usually have combined sewers which convey both the foul sewage, from baths, sinks, wcs, etc., and the surfacewater runoff due to rainfall on paved areas and roofs. Even in moderate rainfalls, the surface water runoff becomes much larger than the d.w.f. from a built-up area and sewers would need to be uneconomically large to contain the flow. It is therefore customary to install storm overflows which divert flows in excess of 6, 9, or sometimes 12, d.w.f. to a nearby water course. This inevitably causes significant pollution although hopefully the flow in the water course receiving the storm discharge will be high because of the rainfall. With combined sewers there are hydraulic design problems related to the need to maintain a minimum self-cleansing velocity at low flows whilst preventing excessive velocities when the sewer is running full. In passing it should be noted that the achievement of self-cleansing velocities in sewers is particularly important in tropical areas since in those conditions, organic deposits will rapidly become anaerobic and the resultant production of hydrogen sulphide can cause serious damage to the sewer. Because of the disadvantages of combined sewers, most new developments are sewered on a separate system with one relatively small foul sewer, all of whose contents are treated, and with storm water sewers which carry only relatively clean runoff and which can be safely discharged to local water courses. The cost of a separate system will inevitably be somewhat higher than that of a combined system although in many cases it may be possible to lay both pipes in the same excavation for at least part of their length.

The rate of surface runoff depends upon the intensity of rainfall and the impermeability of the area drained which varies according to its nature (Table 8.3). Rainfall intensity varies with the return period of the storm and

TABLE 8.3. TYPICAL IMPERMEABILITY FACTORS

Surface	Impermeability factor
Watertight roof	0.70–0.95
Asphalt pavement	0.85–0.90
Concrete flagstones	0.50–0.85
Macadam road	0.25–0.60
Gravel drive	0.15–0.30
Undeveloped land:	
flat	0.10–0.20
sloping	0.20–0.40
steep rocky slope	0.60–0.80

the duration for which it lasts and various empirical relationships for rainfall intensity are available for different geographical locations. The Bilham formula is commonly used in the UK:

$$\text{Rainfall intensity mm/h} = \left[\frac{14.2\, F^{0.28}}{t^{0.72}} - \frac{2.54}{t} \right] \qquad (8.3)$$

where F = return period in years,
       t = storm duration in hours.

The maximum runoff from an area is obtained when the duration of the storm is equivalent to the time of concentration of the area. The time of concentration has two components, the time of entry, i.e. the time for rain to flow along gutters and drainpipes into the sewer, and the time of flow along the sewer.

The runoff from a drainage area is given by:

$$Q \text{ m}^3/\text{s} = 0.278\, ApR \qquad (8.4)$$

where Ap = impermeable area km^2,
       R = rainfall intensity mm/h.

Various design methods for surface-water sewers are described in the literature.[3,4]

8.4. Variations in Flow

Although the average water demand and sewage d.w.f. may be determined for a community as say 150 l/person day there will be considerable variations over a 24-h period. The magnitude of these variations depends upon the size of the population concerned. In the case of water consumption, the ratio of peak hourly rate to annual average rate can vary from about 3 for a population of a few hundred to about 2.2 for a community of 50 000 and about 1.9 for a population of half a million. Similar variations are to be expected with the d.w.f. in sewers and the effects of surface runoff will of course greatly amplify the peak flows is combined systems. Even in separate systems, flows can be affected by rainfall since illegal gulley connections and small areas of impermeable surfaces are often served by the foul sewer.

Water-treatment plants can often operate at a constant rate with the distribution system and service reservoirs serving to balance the fluctuating demand. However, differential electricity tariffs may encourage higher rates of treatment during the night. In the case of sewage-treatment plants the flow of sewage arriving at the works will be balanced to some extent within the sewerage system but the plant must be designed to operate under fluctuating flows with a normal maximum capacity in the main units of 3 d.w.f.

References

1. THACKRAY, J. E., COCKER, V. and ARCHIBALD, G. G., The Malvern and Mansfield studies of domestic water usage. *Proc. Instn Civ. Engrs*, **64**, 1978 (1), 37.
2. THACKRAY, J. E. and ARCHIBALD, G. G., The Severn–Trent studies of industrial water use. *Proc. Instn Civ. Engrs*, **70**, 1981, (1), 403.
3. BARTLETT, R. E., *Public Health Engineering: Sewerage*, 2nd edn, Applied Science Publishers Ltd., Barking, 1978.
4. WHITE, J. B., *Wastewater Engineering*, Edward Arnold, London, 1978.

Further Reading

BARTLETT, R. E. (ed.), *Developments in Sewerage*, **1**, Applied Science Publishers Ltd., Barking, 1979.
CENTRAL WATER PLANNING UNIT, *Analysis of Trends in Public Water Supply*, CWPU, Reading, 1976.
MALES, D. B., *Household Use of Water*, CWPU, Reading, 1975.
MALES, D. B. and TURTON, P. S., *Design Flow Criteria in Sewers and Water Mains*, CWPU, Reading, 1979.
PHILLIPS, J. H. and KERSHAW, C. G., Domestic metering—an engineering and economic appraisal. *J. Instn Wat. Engrs Scits*, **30**, 1976, 203.
STERLING, M. J. H. and ANTCLIFFE, D. J., A technique for prediction of water demand from past consumption data. *J. Instn Wat. Engrs*, **28**, 1974, 413.

Problems

1. Using the census data below for two communities predict the populations in 2001 using arithmetic, geometric and graphical techniques.

Year	Developed-country town	Developing-country city
1921	64 126	257 295
1931	67 697	400 075
1941	69 850	632 136
1951	74 024	789 400
1961	75 321	1 227 996
1971	75 102	2 022 577
1981	73 986	2 876 309

(There are no correct answers for this problem and those obtained will depend on how much of the existing data is utilized.)

2. Determine the rainfall intensity as given by the Bilham formula for storms of 15-min duration with return periods of 1, 5 and 10 years.   (28.4, 50.4 and 63.5 mm/h)

3. Using the rainfall intensities determined above calculate the corresponding runoff values for an area of 3 km^2 which has an average impermeability factor of 0.45.   (10.7, 18.9 and 23.8 m^3/s)

CHAPTER 9

Introduction to Treatment Processes

It will be apparent from previous chapters that waters and wastewaters often have highly complex compositions and that modifications to the composition are usually necessary to suit a particular use. It follows that a variety of treatment processes will be necessary to deal with the range of contaminants likely to be encountered.

Contaminants may be present as:

1. Floating or large suspended solids:
   in water—leaves, branches, etc.;
   in wastewater—paper, rags, grit, etc.
2. Small suspended and colloidal solids:
   in water—clay and silt particles, micro-organisms;
   in wastewater—large organic molecules, soil particles, micro-organisms.
3. Dissolved solids:
   in water—alkalinity, hardness, organic acids;
   in wastewater—organic compounds, inorganic salts.
4. Dissolved gases:
   in water—carbon dioxide, hydrogen sulphide;
   in wastewater—hydrogen sulphide.
5. Immiscible liquids:
   e.g. oils and greases.

The actual particle size at which the nature of the material changes from one group to another depends upon such physical characteristics as specific gravity of the material and the division between groups is in any event indistinct. In certain cases it may be necessary to add substances to the water or wastewater to improve its characteristics, e.g. chlorine for disinfection of water, oxygen for the biological stabilization of organic matter.

9.1. Methods of Treatment

There are three main classes of treatment process:

1. Physical processes which depend essentially on physical properties of the impurity, e.g. particle size, specific gravity, viscosity, etc. Typical examples of this type of process are screening, sedimentation, filtration, gas transfer.
2. Chemical processes which depend on the chemical properties of an impurity or which utilize the chemical properties of added reagents. Examples of chemical processes are: coagulation, precipitation, ion exchange.
3. Biological processes which utilize biochemical reactions to remove soluble or colloidal impurities, usually organics. Aerobic biological processes include biological filtration and activated sludge. Anaerobic oxidation processes are used for the stabilization of organic sludges and high strength organic wastes.

Figure 9.1 shows typical operational ranges of various treatment processes.

In some situations, a single treatment process may provide the desired change in composition but in most cases it is necessary to utilize several processes in combination. For example, sedimentation of a river water will remove some, but by no means all, of the suspended matter. The addition of a chemical coagulant followed by gentle stirring (flocculation) will cause the

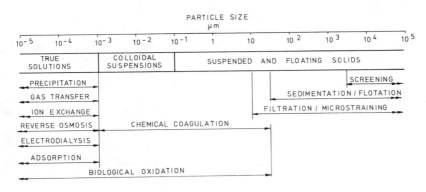

FIG. 9.1. Applications of the main treatment processes.

TABLE 9.1. PROBABLE TREATMENT FOR VARIOUS RAW WATERS

Source	Probable treatment	Possible additions
Upland catchment	Screening or microstraining disinfection	Sand filtration Stabilization Colour removal
Lowland river	(i) Screening or microstraining, coagulation, rapid filtration disinfection	Storage Softening Stabilization
	(ii) Screening or microstraining, rapid filtration, slow filtration, disinfection	Adsorption Desalination Nitrate removal
Deep groundwater	Disinfection	Softening Stabilization Iron removal Desalination Nitrate removal

TABLE 9.2. PROBABLE DOMESTIC SEWAGE TREATMENT FOR VARIOUS RECEIVING WATERS

Receiving water	Typical effluent standard		Probable treatment
	BOD	SS	
Open sea	—	—	Screening or comminution
Tidal estuary	150	150	Screening or comminution, primary sedimentation, sludge disposal by dumping on land or at sea
Lowland river	20	30	Screening or comminution, primary sedimentation, aerobic biological oxidation, secondary sedimentation, sludge stabilization, sludge disposal on land or at sea or by incineration
High-quality river	10	10	As for lowland river with addition of tertiary treatment by sand filtration, grass plot irrigation or lagoons

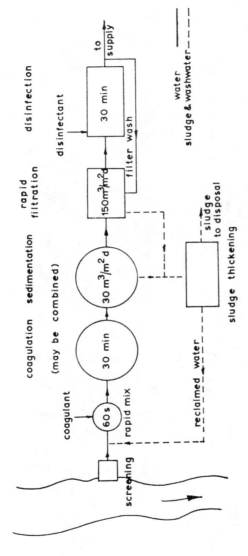

Fig. 9.2. Conventional water-treatment plant.

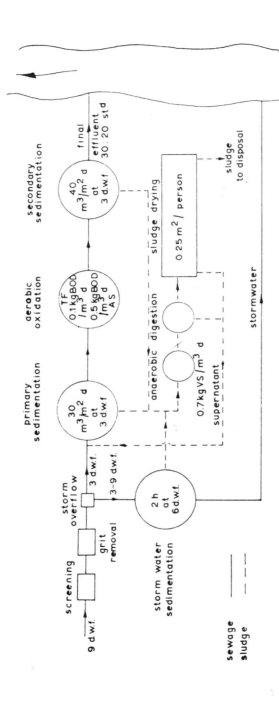

Fig. 9.3. Conventional sewage-treatment plant.

agglomeration of colloidal particles which can then be largely removed by sedimentation. Most remaining non-settleable solids can be removed by filtration through a bed of sand. The addition of a disinfectant serves to kill any harmful micro-organisms which have survived the preceding stages of treatment.

The probable combinations of treatment processes required to produce potable water from various sources are given in Table 9.1 and Table 9.2 shows typical domestic sewage treatment systems for various effluent qualities. Flow sheets and typical design criteria for conventional water-treatment and domestic sewage-treatment plants appear in Figs. 9.2 and 9.3 respectively.

9.2. Optimized Design

As outlined above, treatment plants usually consist of a number of unit processes or operations in combination. Most plants are designed using fairly standard criteria, of the type shown in Figs. 9.2 and 9.3, which have been developed over the years and which will usually produce satisfactory levels of performance. However, it is important to appreciate that such an approach tends to lead to a somewhat conservative approach to design. A more rational approach is based on the concept of the treatment units forming a system in which each unit is designed to perform a particular function and the overall system is optimized economically. Increasing capital and operational costs mean that investments in treatment plants must be carefully scrutinized to ensure that the best value for money is obtained. Whilst it is true that the safe operation of water and wastewater treatment processes can have a highly beneficial effect on public health such activities should not be shielded from rational analysis.

The use of systems analysis concepts to develop mathematical optimizing models of treatment plants can provide a useful aid to the designer provided that reliable performance and cost data are available. A primary requirement for optimization is the availability of performance relationships for each unit process, linking input and output qualities with a characteristic loading parameter. Performance relationships may be established on the basis of knowledge of the theoretical behaviour of the particular process or on the basis of an empirical model for the process. In either case it is necessary to prove that the model developed does provide a satisfactory representation of the process for which it has been produced.

The costs of treatment, both capital and operating, are important factors in any design but the establishment of reliable cost functions is not easy in times of high inflation and escalating energy charges. Wide variations in capital costs are found at different plants since ground conditions or site configuration can have a marked effect on costs and make comparison with data from other plants somewhat difficult. There are, however, clear economies of scale with most treatment facilities so that the *per capita* cost of water or wastewater treatment for a village may be several times greater than for a city.

By combining performance relationships and cost functions it is possible to produce a mathematical model of a complete treatment plant which can be used by a designer to evaluate a number of treatment options and thus arrive at the optimum design.

Further Reading

AMERICAN SOCIETY OF CIVIL ENGINEERS, *Wastewater Treatment Plant Design*, ASCE, New York, 1977.

AMERICAN WATER WORKS ASSOCIATION, *Water Treatment Plant Design*, AWWA, New York, 1969.

BOWDEN, K., GALE, R. S. and WRIGHT, D. E., Evaluation of the CIRIA prototype model for the design of sewage treatment works. *Wat. Pollut. Control*, 75, 1976, 192.

DEGREMONT, *Water Treatment Handbook*, John Wiley & Sons, New York, 1979.

INSTITUTE OF WATER POLLUTION CONTROL, Manuals of British practice in water pollution control.
    Preliminary Processes, 1972.
    Primary Sedimentation, 1973.
    Tertiary Treatment and Advanced Waste Water Treatment, 1974.
    Sewage Sludge 1, 2 and 3, 1978–1981.
    IWPC Maidstone.

METCALF and EDDY, *Wastewater Engineering: Treatment Disposal Reuse*, 2nd edn, Tata McGraw-Hill, New Delhi, 1979.

OKUN, D. A. and PONGHIS, G., *Community Wastewater Collection and Disposal*, WHO, Geneva, 1975.

TEBBUTT, T. H. Y., Developments in performance relationships for sewage treatment. *Pub. Hlth Engr*, 6, 1978, 79.

TWORT, A. C., HOATHER, R. C. and LAW, F. M., *Water Supply*, 2nd edn, Edward Arnold, London, 1974.

CHAPTER 10

Preliminary Treatment Processes

To PROTECT the main units of a treatment plant and to aid in their efficient operation it is necessary to remove the large floating and suspended solids which are often present in the inflow. These materials include leaves, twigs, paper, rags and other debris which could obstruct flow through a plant or damage equipment in the plant.

10.1. Screening and Straining

The first stage in preliminary treatment usually involves a simple screening or straining operation to remove large solids. In the case of water treatment some form of protective boom or coarse screen with openings of about 75 mm is used to prevent large objects reaching the intake. The main screens are usually provided in the form of a mesh with openings of 5–20 mm and arranged as a continuous belt, a disc or a drum through which the flow must pass (Fig. 10.1). The screening mesh is usually slowly rotated so that the material collected can be removed before an excessive head loss is reached. The screenings removed from water are normally returned to the source downstream of the abstraction point.

With sewage the content of paper and rags is often high and the nature of the materials is such that a mesh screen would be difficult to keep clean. It is therefore customary to use a bar screen arrangement with a spacing between bars of 20–60 mm. On small works, intermittent hand cleaning of screens is possible but on larger installations automatic mechanical cleaning is provided either on the basis of elapsed time or initiated by the buildup of head loss across the screen. Sewage screenings are unpleasant in nature and are usually disposed of by burial or incineration. Alternatively they may be passed to a macerator which shreds them to a small size so that

103

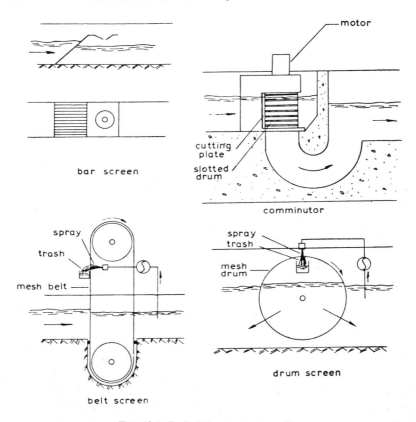

FIG. 10.1. Preliminary treatment units.

they can then be returned to the flow for removal with the rest of the settleable solids during the main treatment process. In some situations the use of
a comminutor which shreds the solids *in situ* may be preferred (Fig. 10.1).
This obviates the need for the handling of solids but requires rather more
head than would be lost through a screen.

10.2. Microstraining

The microstrainer is a development of the drum screen which uses a fine woven stainless-steel mesh with aperture sizes of 20–60 μm to provide removal of relatively small solids. It has applications in water treatment for removal of algae and similar-size particles from waters of otherwise good quality. Microstraining is also employed as a final tertiary stage to produce a high-quality sewage effluent. Because of the small mesh apertures, clogging occurs rapidly so that the drum is rotated at a peripheral speed of about 0.5 m/s and the mesh continually washed clean by high-pressure sprays. Straining rates in normal usage are 750–2500 m^3/m^2 d. The design of microstrainer installations is based on the laboratory determination of an empirical characteristic of the suspension known as the filtrability index. This parameter measures the behaviour of the suspension with reference to its clogging properties and can be used to determine the allowable straining rate to prevent excessive clogging and possible physical damage to the mesh.

10.3. Grit Removal

In most sewerage systems and particularly those with combined sewers, considerable amounts of grit are carried along in the flow and this material if not removed could cause damage to mechanical parts of the treatment plant. Because the grit particles are relatively large, with a high density compared with the organic particles in sewage, they are often removed using the principle of differential settling. Grit particles with a diameter of 0.20 mm and S.G. 2.65 have a settling velocity of about 1.2 m/min whereas most of the suspended solids in sewage have considerably lower settling velocities. By using a parabolic section channel it is possible to provide a constant horizontal velocity of around 0.3 m/s at all rates of flow. Under these conditions a channel of sufficient length to provide a retention time of 30 – 60 seconds will allow the grit particles to settle to the bottom whilst the remaining suspended solids are still transported by the flow. The grit is removed at intervals, washed and then disposed of for re-use in some way. Other types of grit-removal device may involve an aerated spiral-flow chamber to achieve the desired separation or the use of a short retention settling tank, any organic solids removed with the grit being washed back into the flow before the grit is discharged.

10.4. Flow Distribution

In most treatment plants it is necessary to divide the flow between a number of similar units or to discharge flows in excess of a design maximum to supplementary units, e.g. stormwater settling tanks. Such flow division is not easy to provide in a satisfactory manner since the hydraulics of treatment plant are often complex. Probably the most suitable form of flow division is achieved by free-fall weirs although in some cases the head required for such structures may not be readily available.

Further Reading

BOUCHER, P. L., Micro-straining. *J. Instn Pub. Hlth Engrs*, **60**, 1961, 294.
INSTITUTE OF WATER POLLUTION CONTROL, *Preliminary Processes*, IWPC, Maidstone, 1972.
ROEBUCK, I. H. and GRAHAM, N. J. D., Pilot plant studies of fine screening in raw sewage. *Pub. Hlth Engr*, **8**, 1980, 154.
WHITE, J. B., Aspects of the hydraulic design of sewage treatment works. *Pub. Hlth Engr*, **10**, 1982, 164.

CHAPTER 11

Clarification

MANY of the impurities in water and wastewater occur as suspended matter which remains in suspension in flowing liquids but which will move vertically under the influence of gravity in quiescent or semi-quiescent conditions. Usually the particles are denser than the surrounding liquid so that sedimentation takes place but with very small particles and with low-density particles flotation may offer a more satisfactory clarification process. Sedimentation units have a dual role—the removal of settleable solids and the concentration of the removed solids into a smaller volume of sludge.

11.1. Theory of Sedimentation

In sedimentation it is necessary to differentiate between discrete particles which do not change in size, shape or mass, during settling, and flocculent particles which agglomerate during settling and thus do not have constant characteristics.

The basic theory of sedimentation assumes the presence of discrete particles. When such a particle is placed in a liquid of lower density it will accelerate until a limiting terminal velocity is reached, then:

$$\text{gravitational force} = \text{frictional drag force} \tag{11.1}$$

Now
$$\text{gravitational force} = (\rho_s - \rho_w)gV \tag{11.2}$$

where ρ_s = density of particle,
ρ_w = density of fluid,
V = volume of particle.

107

By dimensional analysis it can be shown that

$$\text{frictional drag force} = C_D A_c \rho_w \frac{v_s^2}{2} \tag{11.3}$$

where C_D = Newton's drag coefficient,
A_c = cross-sectional area of particle,
v_s = settling velocity of particle.
C_D is not constant, but varies with Reynolds Number (R) and, to a lesser extent, with the shape of the particle.
For spheres

$$R \leqslant 0.5 \qquad\qquad C_D = \frac{24}{R} \tag{11.4}$$

$$0.5 < R \leqslant 10^4 \qquad\qquad C_D = \frac{24}{R} + \frac{3}{\sqrt{R}} + 0.34 \tag{11.5}$$

In sedimentation $\qquad\qquad R = \frac{v_s d}{v} \tag{11.6}$

where d = particle diameter,
v = kinematic viscosity of the fluid.
Equating gravitational and frictional forces for the equilibrium condition

$$(\rho_s - \rho_w) g V = C_D A_c \rho_w \frac{v_s^2}{2} \tag{11.7}$$

i.e. $\qquad\qquad v_s = \sqrt{\frac{2g V (\rho_s - \rho_w)}{C_D \rho_w A_c}} \tag{11.8}$

for spheres $\quad V = \frac{\pi d^3}{6} \quad A_c = \frac{\pi d^2}{4}$

hence $\qquad\qquad v_s = \sqrt{\frac{4g d (\rho_s - \rho_w)}{3 C_D \rho_w}} \tag{11.9}$

or $\qquad\qquad v_s = \sqrt{\frac{4g d}{3 C_D} (S_s - 1)} \tag{11.10}$

where S_s = specific gravity of the particle.

For turbulent flow $\quad 5 \times 10^2 < R < 10^4$

$\qquad C_D$ tends to 0.4

Thus $\qquad\qquad v_s = \sqrt{3.3gd(S_s - 1)}$ (11.11)

For laminar flow $\quad R \leqslant 0.5$

$$C_D = \frac{24}{R}$$

Thus $\qquad\qquad v_s = \frac{gd^2(S_s - 1)}{18v}$ (11.12)

which is Stokes's Law.

In calculating settling velocities it is essential to check that the correct formula (11.11) or (11.12) has been used for the velocity as determined. In the transitional range between turbulent and laminar flow a trial and error solution for v_s may be used.

When dealing with flocculent suspensions it is not possible to apply the above theory because the agglomeration of floc particles results in increased settling velocity with depth due to the formation of larger and heavier particles; this feature is illustrated in Fig. 11.1. Many of the suspensions in the treatment of water and wastewater are flocculent in nature.

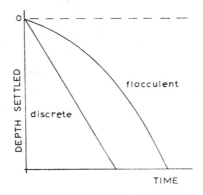

FIG. 11.1. Settlement of discrete and flocculent particles.

Four different types of settling can occur:

Class 1 *Settling*: settlement of discrete particles in accordance with theory.

Class 2 *Settling*: settlement of flocculent particles exhibiting increased velocity during the process.

Zone Settling: at certain concentrations of flocculent particles the particles are close enough together for the interparticulate forces to hold the particles fixed relative to one another so that the suspension settles as a unit.

Compressive Settling: at high concentrations the particles are in contact and the weight of the particles is in part supported by the lower layers of solids.

In the case of concentrated suspensions (> 2000 mg/l SS) hindered settlement occurs. In these circumstances there is a significant upward displacement of water due to the settling particles and this has the effect of reducing the apparent settling velocity of the particles (Fig. 11.2).

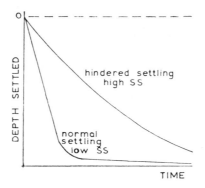

FIG. 11.2. Hindered settling.

11.2. The Ideal Sedimentation Basin

The behaviour of a sedimentation tank operating on a continuous flow basis with a discrete suspension of particles can be examined by reference to an ideal sedimentation basin (Fig. 11.3) which assumes:

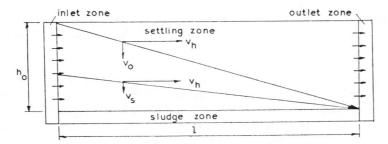

FIG. 11.3. The ideal sedimentation basin.

1. Quiescent conditions in the settling zone.
2. Uniform flow across the settling zone.
3. Uniform solids concentration as flow enters the settling zone.
4. Solids entering the sludge zone are not resuspended.

Considering a discrete particle with a settling velocity v_0 which just enters the sludge zone at the end of the tank. This particle falls through a depth h_0 in the retention time of the tank t_0:

$$v_0 = \frac{h_0}{t_0} \qquad (11.13)$$

But since

$$t_0 = \frac{\text{Volume } (V)}{\text{Flow/unit time } (Q)}$$

$$v_0 = \frac{h_0 \times Q}{V} = \frac{h_0 \times Q}{A \times h_0} \qquad (11.14)$$

where A = surface area of tank

$$v_0 = \frac{Q}{A} \qquad (11.15)$$

Q/A is termed the surface overflow rate and it follows from (11.15) that for discrete particles solids removal is not dependent on the depth of the tank. For flocculent particles, however, depth does affect solids removal since the deeper the tank the more likely it is that agglomeration will occur and hence a larger proportion of the solids would be removed.

If a tank is fed with a suspension of discrete particles of varying sizes it is possible to determine the overall removal as follows, again referring to Fig. 11.3.

The tank is designed to remove all particles with settling velocity $\geqslant v_0$. Particles with settling velocity $v_s < v_0$ will only be removed if they enter the tank at a distance from the bottom not greater than h where $h = v_s t_0$. Thus the proportion of particles with $v_s < v_0$ which will be removed is given by (v_s/v_0). By inserting a series of trays in the tank at a spacing of $v_s t_0$ it would theoretically be possible to remove all settleable solids. High-rate tube or inclined plate settling units use this principle to improve removal efficiencies.

11.3. Measurement of Settling Characteristics

The settling velocity of individual particles may be determined by timing their fall through a known depth of fluid, but for graded suspensions a settling column analysis is more useful. A settling column is a tube 2–3 m deep with tapping points at intervals, the diameter of the tube being at least a hundred times the largest particle size to prevent wall effects.

For an analysis the suspension is thoroughly mixed in the column and the initial SS concentration determined as c_0 mg/l. A sample is taken at depth h_1 after time t_1 and the SS concentration is found to be c_1 mg/l. Now all particles with a settling velocity greater than $v_1 (= h_1/t_1)$ will have settled past the sampling point and the particles remaining, i.e. c_1 must have a settling velocity less than v_1. Thus the proportion of particles p_1 have a settling velocity less than v_1 is given by

$$p_1 = \frac{c_1}{c_0} \qquad (11.16)$$

The procedure is repeated for time intervals t_2, t_3, . . . , and hence proportions of particles p_2, p_3, . . . , having settling velocities less than v_2, v_3, . . . , are determined. Plotting these data gives the settling characteristic curve for the suspension (Fig. 11.4).

In a tank with an overflow velocity of v all particles with $v_s > v$ will be removed regardless of the position at which they enter the tank. In addition, for a horizontal flow tank, particles with $v_s < v$ will be removed if they enter

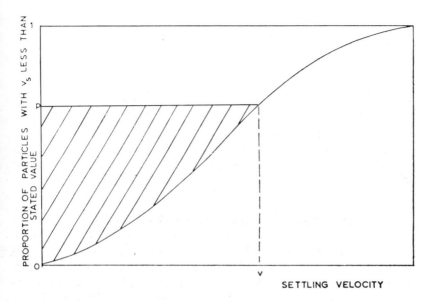

FIG. 11.4. Settling characteristic curve for a discrete suspension.

at a distance from the bottom not exceeding $v_s t_0$. Thus from Fig. 11.4 the overall removal in a horizontal flow tank is given by

$$P = 1 - p + \int_0^p \frac{v_s}{v} dp \qquad (11.17)$$

Figure 11.5 gives a diagrammatic illustration of this point.

For a vertical flow tank

$$P = 1 - p \qquad (11.18)$$

since all particles with $v_s < v$ will eventually be washed out of the tank.

When dealing with flocculent suspensions both concentration of particles and the effect of depth on flocculation must be taken into account. Samples are taken from a number of depths at each time interval and the SS in each sample are expressed as a percentage of the original concentration. The

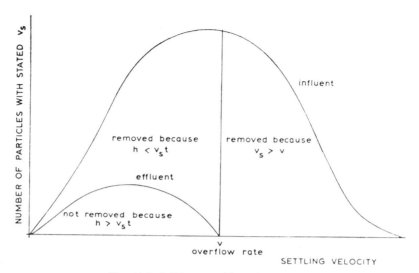

FIG. 11.5. Solids removal by sedimentation.

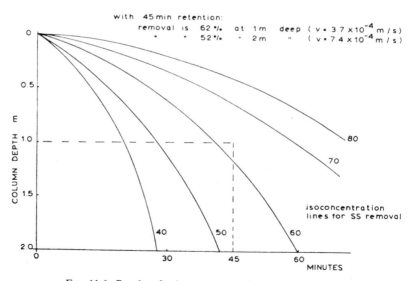

FIG. 11.6. Results of column test on a flocculent suspension.

difference between this percentage and 100 is thus the percentage of solids which have settled past the sampling point and would therefore have been removed in a tank of that particular depth and retention time. Plotting these results as in Fig. 11.6 enables the construction of smooth iso-concentration lines for the SS removal at various depth and time conditions. The effect of flocculation is shown in the shape of the iso-concentration lines, the greater the slopes of the lines with depth the greater the degree of flocculation which has taken place. The performance of alternative tanks can be predicted from such a plot and a particular unit selected after consideration of performance and cost.

11.4. Efficiency of Sedimentation Tanks

The hydraulic behaviour of a tank may be examined by injecting a tracer into the inlet and observing its appearance in the effluent. The flow-through curves so obtained are of infinite variety, ranging from the ideal plug-flow case to that of a completely mixed tank as shown in Fig. 11.7. The flow-through curve obtained in practice is a combination of the two extremes, short-circuiting due to density currents and mixing due to hydraulic turbulence producing a peak earlier than would be expected in an ideal tank. Thus the actual retention time is often considerably less than the theoretical value.

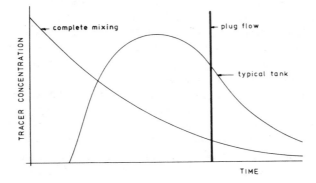

FIG. 11.7. Flow-through curves.

Since the purpose of sedimentation tanks is to remove suspended matter the logical way of expressing their efficiency is by the percentage removal of such solids. The normal SS determination records particles down to a few microns whereas floc particles smaller than $100\,\mu m$ are unlikely to be removed by sedimentation. Thus a sedimentation tank will never remove all the SS from sewage and the normal range of SS removal from sewage by sedimentation is 50–60%. Research has shown that with heterogeneous suspensions such as sewage the hydraulic loading on a tank has less influence on the removal efficiency than the influent SS concentration.

11.5. Types of Sedimentation Tank

The main types of sedimentation tank found in practice are shown in Fig. 11.8. The horizontal tank is compact but suffers from a restricted effluent weir length unless suspended weirs are adopted. Sludge is moved to the sump by a travelling bridge scraper which may serve several tanks or by a continuous-belt system with flights. The sludge is withdrawn from the sump under hydrostatic head. Circular tanks offer advantages of long weir length and simpler scraping mechanisms but are not so compact. Hopper-bottom tanks with horizontal flow are popular on small sewage works where the extra construction cost is more than offset by the absence of any scraping mechanism. The vertical-flow hopper-bottom tank is often used in water-treatment plants and in such conditions operates with a sludge blanket which serves to strain out particles smaller than would be removed by sedimentation alone at the overflow rate employed.

Many different designs of inlet and outlet structures are in use and whilst particular designs may offer some improvement in solids removal with homogeneous flocculent suspensions they usually make little difference to the removal of SS from raw sewage.

Sedimentation tanks have two functions: the removal of settleable solids to produce an acceptable output and the concentration of the removed solids into a smaller volume. The design of a tank must consider both of these functions and the tank should be sized on whichever of the requirements is limiting. The sludge thickening function of a tank is likely to be important when dealing with relatively high concentrations of homogeneous solids.

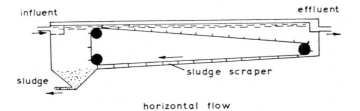

horizontal flow

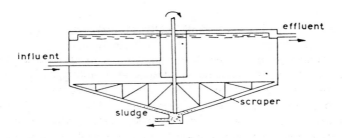

radial flow

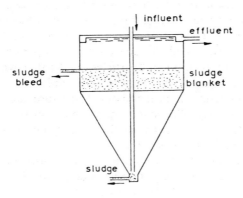

vertical flow

FIG. 11.8. Types of sedimentation tank.

11.6. Gravity Thickening

The design of sedimentation tanks must take into account both the solids removal and the sludge-thickening functions. The size of a unit will be limited by one of these functions and in the case of high concentrations of homogeneous suspensions like activated-sludge flocs or chemical precipitation flocs the thickening function may be the more critical.

Analogous to the ideal settling-basin concept it is possible to envisage an ideal thickener which has uniform horizontal distribution of particles and from the base of which the thickened suspension is removed creating equal downward velocity across the tank section. It is also assumed that an ideal suspension having incompressible solids is present. As shown in Fig. 11.9 the flux of solids past a point in the thickener is the result of the rate of downward movement under gravity and the rate of downward movement due to the removal of solids by withdrawal of sludge. The solids flux G_c is given by

$$G_c = c_i v_i + c_i v_w \qquad (11.19)$$

where c_i = solids concentration,
$\quad\quad v_i$ = settling velocity of solids at concentration c_i,
$\quad\quad v_w$ = downward velocity produced by withdrawal.

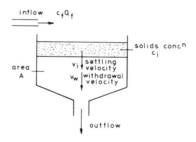

FIG. 11.9. Gravity thickening.

The batch settlement flux $(c_i v_i)$ is governed by the physical characteristics of the suspension. The withdrawal flux $(c_i v_w)$ is an operational effect directly proportional to the sludge-removal rate. Figure 11.10 shows how the two fluxes change as the solids concentration increases. The reason for the shape

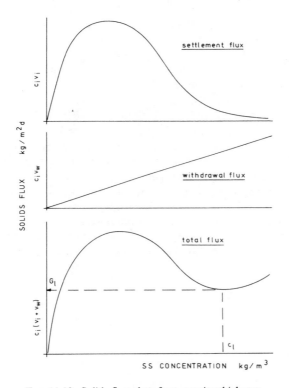

FIG. 11.10. Solids flux plots for a gravity thickener.

of the settling flux plot is that at low SS levels the settling velocity will be high but as the SS level increases hindered settling effects cause a progressive reduction in the bulk settling velocity. The total flux curve can then be used to aid in the design of the thickening function of clarifiers. Considering the suspension whose characteristics are shown in the lower portion of Fig. 11.10, it is clear that the concentration c_l corresponds to a minimum total flux level so that to thicken the sludge to a concentration of c_l or higher it will be necessary to ensure that the applied solids load does not exceed G_l,

$$\text{i.e. Applied load} = \frac{c_f Q_f}{A} \leqslant G_l \qquad (11.20)$$

where c_f = solids concentration in tank feed,
  Q_f = flow rate to tank,
  A = cross-sectional area of thickener.

If the solids flux is higher than G_l not all of the solids would be able to reach the sludge outlet. To remove all solids at a flux greater than G_l it would be necessary either to increase the area of the tank or to increase the removal velocity which would give reduced thickening since the solids concentration in the withdrawn flow is inversely proportional to the removal velocity.

It is possible to obtain this type of information about the thickening characteristics of suspensions directly from the batch settlement flux curve alone. It can be shown that if a tangent is drawn to the settlement flux curve at the rate-limiting concentration the slope of the tangent will be the necessary downward velocity in a continuous thickener to maintain a constant sludge level. The intercept on the y-axis will be the solids flux for the equilibrium condition and the intercept on the x-axis will be the solids concentration in the withdrawn sludge. Figure 11.11 shows the use of the

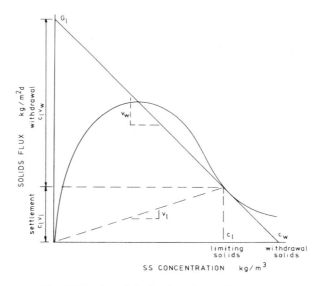

FIG. 11.11. Use of the batch settlement flux curve.

settlement curve in this manner. The slope of the line from the origin to the point of tangency is the gravitational settling velocity at c_l and since the slope of the tangent is v_w, the intercept is G_l and the total flux made up from the two components as shown. This type of plot may be used to determine the required thickener area for various thickened sludge concentrations. Similarly the plot can be used to predict the effect of a change in loading on an existing tank with respect to sludge concentration and the necessary withdrawal rate.

11.7. Flotation

An alternative clarification technique, which is particularly attractive for relatively small particles and for particles with a density close to that of water, is flotation. With flotation the loading rates are not directly related to the suspension characteristics so it is usually possible to provide relatively short retention times whilst still obtaining good clarification. The process involves the addition of a flotation agent, usually fine air bubbles, which becomes associated with the suspended particles and thus provides the necessary buoyancy to carry them to the surface of the tank where they can be removed as a scum. Air flotation requires the release of a cloud of fine air bubbles at the base of the unit and this is usually achieved by saturating a portion of the treated flow (the recycle) with air at high pressure. When this pressurized liquid is returned to the main flow at atmospheric pressure the excess air comes out of solution in the desired fine-bubble form. The bubbles of air become attached to or enmeshed in the suspended particles which then rise to the surface because of their reduced density.

Figure 11.12 shows that schematic arrangement of a typical dissolved air flotation unit. For water-treatment operations recycle ratios of around 10% with pressurization up to $400\,\mathrm{kPa}$ have proved satisfactory, giving rise rates of about $12\,\mathrm{m/h}$ with good clarification. The scum removed from the tank surface usually has a significantly higher solids content than that achievable by sedimentation of the same suspension. The capital cost of flotation units is less than that of the equivalent sedimentation units but operating costs are higher so that the main application of flotation is with suspensions having relatively low settling velocities.

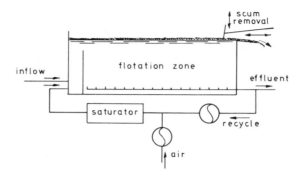

FIG. 11.12. Dissolved air flotation.

Further Reading

BRADLEY, R. M., The operating efficiency of circular primary sedimentation tanks in Brazil and the United Kingdom. *Pub. Hlth Engr*, 1975 (13), Jan. 5.

DICK, R. I., Gravity thickening of sewage sludges. *Wat. Pollut. Control*, **71**, 1972, 368.

HAZEN, A., On sedimentation. *Trans. Am. Soc. Civ. Engrs*, **53**, 1904, 45. (Reprinted *J. Proc. Inst. Sew. Purif.*, 1961 (6), 521.)

KYNCH, G. J., A theory of sedimentation. *Trans. Faraday Soc.* **48**, 1952, 166.

MARCH, R. P. and HAMLIN, M. J., An investigation into the performance of a full-scale sedimentation tank. *J. Proc. Inst. Sew. Purif.*, 1966 (2), 118.

MORSE, J. J., Dissolved air flotation in water treatment. *Wat. Wat. Engng*, **77**, 1973, 161.

PRICE, G. A. and CLEMENTS, M. S., Some lessons from model and full-scale tests in rectangular sedimentation tanks. *Wat. Pollut. Control*, **73**, 1974, 102.

TEBBUTT, T. H. Y., The performance of circular sedimentation tanks. *Wat. Pollut. Control*, **68**, 1969 (4), 467.

TEBBUTT, T. H. Y., Primary sedimentation of wastewater. *J. Wat. Pollut. Control Fedn*, **51**, 1979, 2858.

TEBBUTT, T. H. Y. and CHRISTOULAS, D. G., Performance relationships for primary sedimentation. *Wat. Res.* **9**, 1975, 347.

WILKINSON, P. D., BOLAS, P. M. and ADKINS, M. F., Bewl Bridge Treatment Works. *J. Inst. Wat. Engrs Scits*, **35**, 1981, 47.

WILLIS, R. M., Tubular Settlers—a technical review. *J. Am. Wat. Wks Assn*, **70**, 1978, 331.

YAO, K. M., Design of high-rate settlers. *J. Envr. Engng Div. Am. Soc. Civ. Engrs*, **99**, 1973, 621.

Problems

1. Find the settling velocity of spherical discrete particles 0.06 mm diameter S.G. 2.5 in water at 20°C ($v = 1.010 \times 10^{-6}$ m^2/s). (0.0029 m/s)

2. A settling tank is designed to remove spherical discrete particles 0.5 mm diameter S.G. 1.01 from water at 20°C. Assuming ideal settling conditions, determine the removal of spherical discrete particles 0.2 mm diameter S.G. 1.01 by this tank. (16%)

3. Settling column tests on a discrete particle suspension gave the following results from a depth of 1.3 m:

Sampling time, min	5	10	20	40	60	80
% of initial SS in sample	56	48	37	19	5	2

Determine the theoretical removal of solids from this suspension in a horizontal flow tank with surface overflow rate of 200 m^3/m^2 d. (67%)

4. Tests on a flocculent suspension in a settling column with three sampling points gave the following results:

Sample time, min	% SS removed at		
	1 m	2 m	3 m
0	0	0	0
10	30	18	16
20	60	48	40
30	62	61	60
40	70	63	61
60	73	69	65

Determine the probable removal of solids from this suspension in a tank 2 m deep with alternative retention times of 25 and 50 min. (53%, 65%)

5. Laboratory studies on a floc suspension produced the following settling data:

SS mg/l	2500	5000	7500	10 000	12 500	15 000	17 500	20 000
Settling velocity mm/s	0.80	0.41	0.22	0.10	0.04	0.02	0.01	0.01

If the suspension is to be thickened to a concentration of 2% (20 000 mg/l) determine the thickener cross-sectional area required for a flow of 5000 m^3/d with an initial SS of 3000 mg/l. If a thickened sludge solids content of 1.5% was acceptable determine the new cross-sectional area required. Determine the minimum cross-sectional area required for the settling function of the tank if the settling velocity is assumed to be 0.8 mm/s in this region of the tank. (148 m^2, 63 m^2, 72.3 m^2)

CHAPTER 12

Coagulation

MANY impurities in water and wastewater are present as colloidal solids which will not settle. Their removal can be achieved by promoting agglomeration of such particles by flocculation with or without the use of a coagulant followed by sedimentation or flotation.

12.1. Colloidal Suspensions

Sedimentation can be used to remove suspended particles down to a size of about 50 μm depending on their density, but smaller particles have very low settling velocities so that removal by sedimentation is not feasible. Table 12.1 gives calculated settling velocities for particles with S.G. 2.65 in water at 10°C. It can be seen that the smaller particles have virtually non-existent settling velocities. If these colloidal particles can be persuaded to agglomerate they may eventually increase in size to such a point that removal by sedimentation becomes possible. In a quiescent liquid fine particles collide because of Brownian movement and also when rapidly settling solids overtake more slowly settling particles. As a result larger

TABLE 12.1. SETTLING VELOCITIES FOR PARTICLES
OF S.G. 2.65 IN WATER AT 10°C

Particle size μm	Settling velocity m/h
1000	6×10^2
100	2×10^1
10	3×10^{-1}
1	3×10^{-3}
0.1	1×10^{-5}
0.01	2×10^{-7}

particles, fewer in number, are produced; growth by these means is, however, slow. Collisions between particles can be improved by gentle agitation, the process of flocculation, which may be sufficient to produce settleable solids from a high concentration of colloidal particles. With low concentrations of colloids a coagulant is added to produce bulky floc particles which enmesh the colloidal solids.

12.2. Flocculation

Agitation of water by hydraulic or mechanical mixing causes velocity gradients the intensity of which controls the degree of flocculation produced. The number of collisions between particles is directly related to the velocity gradient and it is possible to determine the power input required to give a particular degree of flocculation as specified by the velocity gradient.

Consider an element of fluid undergoing flocculation (Fig. 12.1). The element will be in shear and thus

$$\text{power input} = \tau \Delta_x \Delta_z \Delta_y \frac{dv}{dy} \tag{12.1}$$

where τ = shear stress.

$$\text{Power unit volume} = P = \tau \frac{\Delta_x \Delta_y \Delta_z}{\Delta_x \Delta_y \Delta_z} \frac{dv}{dy} \tag{12.2}$$

$$= \tau \frac{dv}{dy} \tag{12.3}$$

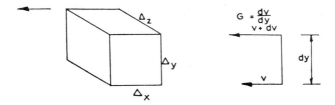

FIG. 12.1. Fluid particle undergoing flocculation.

but by definition $\tau = \mu \dfrac{dv}{dy}$

where μ = absolute viscosity.

Thus
$$P = \mu \frac{dv}{dy} \cdot \frac{dv}{dy} = \mu \left(\frac{dv}{dy} \right)^2 \qquad (12.4)$$

putting
$$G = \frac{dv}{dy}$$

$$P = \mu G^2 \qquad (12.5)$$

For hydraulic turbulence in a baffled tank

$$P = \frac{\rho g h}{t} \qquad (12.6)$$

where ρ = mass density of the fluid,
h = head loss in tank,
t = retention time of tank.

Now from (12.5)
$$G = \sqrt{\frac{P}{\mu}}$$

$$G = \sqrt{\frac{\rho g h}{\mu t}} = \sqrt{\frac{g h}{v t}} \qquad (12.7)$$

In the case of a mechanically stirred tank

$$P = \frac{D v}{V} \qquad (12.8)$$

where D = drag force on paddles,
v = velocity of paddles relative to fluid (usually about three-quarters of blade velocity),
V = volume of tank.

From equation (11.3)
$$D = C_D A \rho \frac{v^2}{2} \qquad (12.9)$$

Thus
$$P = \frac{C_D A \rho v^3}{2V} \qquad (12.10)$$

Substituting for P from (12.5)

$$G^2 = \frac{C_D A \rho v^3}{2 V \mu} \tag{12.11}$$

i.e.

$$G = \sqrt{\frac{C_D A v^3}{2 v V}} \tag{12.12}$$

For good flocculation G should be in the range 20–70 m/s m. Lower values will give inadequate flocculation and higher values will tend to shear the floc particles. The normal retention time in flocculation tanks is 30–45 min. With mechanical flocculation the tank depth is usually $1\frac{1}{2}$–2 paddle diameters and the blade area is 10–25 % of the tank cross-sectional area.

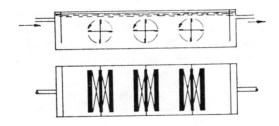

rectangular stirred

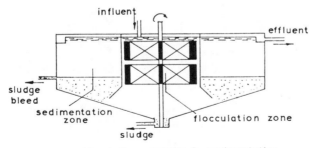

combined flocculation & sedimentation

FIG. 12.2. Flocculation tanks.

Mechanical flocculators provide more control over the process than hydraulic flocculators but require more maintenance. Flocculation and sedimentation may be combined in a single unit (Fig. 12.2) and the sludge blanket type of tank is very popular for water treatment purposes.

12.3. Coagulation

Flocculation of dilute colloidal suspensions provides only infrequent collisions and agglomeration does not occur to any marked extent. In such circumstances clarification is best achieved using a chemical coagulant followed by flocculation and sedimentation. Before flocculation can take place it is essential to disperse the coagulant, usually required in doses of 30–100 mg/l, throughout the body of water. This is carried out in a rapid mixing chamber with a high-speed turbine (Fig. 12.3) or by adding the coagulant at a point of hydraulic turbulence, e.g. at a hydraulic jump in a measuring flume. The coagulant is a metal salt which reacts with alkalinity in the water to produce an insoluble metal hydroxide floc which incorporates the colloidal particles. This fine precipitate is then flocculated to produce settleable solids.

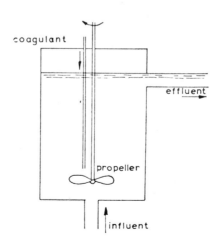

FIG. 12.3. Rapid mixer.

The most popular coagulant for water treatment is aluminium sulphate (alum) $Al_2(SO_4)_3$ and the complex reactions which take place following its addition to water are often simplified as:

$$Al_2(SO_4)_3 + 6H_2O \rightarrow \underline{2Al(OH)_3} + 3H_2SO_4$$
$$3H_2SO_4 + 3Ca(HCO_3)_2 \rightarrow 3CaSO_4 + 6H_2CO_3$$
$$6H_2CO_3 \rightarrow 6CO_2 + 6H_2O$$

i.e. overall

$$Al_2(SO_4)_3 + 3Ca(HCO_3)_2 \rightarrow \underline{2Al(OH)_3} + 3CaSO_4 + 6CO_2$$

When using commercial alum $Al_2(SO_4)_3 \cdot 14H_2O$ it is found that

1 mg/l alum destroys 0.5 mg/l alkalinity $\Big\}$ both as $CaCO_3$
    produces 0.44 mg/l carbon dioxide

Thus for satisfactory coagulation sufficient alkalinity must be available to react with the alum and also to leave a suitable residual in the treated water.

The solubility of $Al(OH)_3$ is pH dependent and is low between pH 5 and 7.5; outside this range coagulation with aluminium salts is not successful. Other coagulants sometimes used are

ferrous sulphate (copperas) $FeSO_4 \cdot 7H_2O$
ferric sulphate $Fe_2(SO_4)_3$
ferric chloride $FeCl_3$

Copperas is sometimes treated with chlorine to give a mixture of ferric sulphate and ferric chloride known as chlorinated copperas. Ferric salts give satisfactory coagulation above pH 4.5, but ferrous salts are only suitable above pH 9.5. Iron salts are cheaper than alum but unless precipitation is complete residual iron in solution can be troublesome particularly due to its stain-producing properties in washing machines.

With very low concentrations of colloidal matter floc formation is difficult and coagulant aids may be required. These may be simple additives like clay particles which form nuclei for precipitation of the hydroxide or polyelectrolytes, heavy long-chain synthetic polymers, which added in small amounts (< 1 mg/l) promote agglomeration and toughen the floc. Because of the spongy nature of floc particles they have a very large surface area and are thus capable of adsorption of dissolved matter from solution. This surface-active effect results in coagulation removing dissolved colour as well as colloidal turbidity from water; a sample of raw water with colour 60°H and turbidity 30 units would usually be improved to about 5°H and 5 units of turbidity after coagulation, flocculation and sedimentation.

It is not possible to calculate the dose of coagulant required nor the results which it will produce so that laboratory tests must be carried out using the jar-test procedure. This involves setting up a series of samples of water on a special multiple stirrer rig and dosing the samples with a range of coagulant, e.g. 0, 10, 20, 30, 40, and 50 mg/l stirring vigorously with a glass rod. The samples are then flocculated for 30 min and allowed to stand in quiescent conditions for 60 min. The supernatant water is then examined

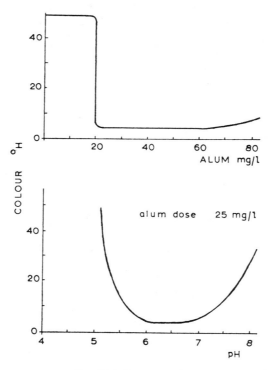

FIG. 12.4. Jar-test results.

for colour and turbidity and the lowest dose of coagulant to give satisfactory removal is noted. A second set of samples is prepared with pH adjusted over a range, for example of 5.0, 6.0, 6.5, 7.0, 7.5, 8.0, and the coagulant dose determined previously added to each beaker followed by stirring, flocculation and settlement as before. It is then possible to examine the supernatant and select the optimum pH and if necessary recheck the minimum coagulant dose required. Figure 12.4 shows typical results from such a jar test. Because of the effect of pH on coagulation it is normally necessary in chemical coagulation plants to make provision for the control of pH by the addition of acid or alkali.

12.4. Mechanism of Coagulation

Although chemical coagulation is a widely used process the mechanisms by which it operates are not fully understood in spite of considerable research effort. Basic colloid stability considerations have been applied to coagulation in attempts to offer explanations for the observed results. The stability of hydrophobic colloid suspensions can be explained by

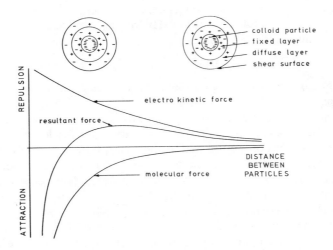

FIG. 12.5. Forces acting on a floc particle.

consideration of the forces acting on the particles as shown in Fig. 12.5. Mutual repulsion arises from the electrostatic surface charges but destabil- ization can be achieved by the addition of ions of opposite charge to reduce the repulsive forces and permit the molecular attraction forces to become dominant. In this context the value of the zeta potential, the electrical potential at the edge of the particle agglomerate, is of some significance. In theory a zero zeta potential should provide the best conditions for coagulation. However, when dealing with the heterogeneous suspensions found in water it seems that there are many complicating factors and zeta potential measurements are not always of much value in operational circumstances.

In the case of relatively low suspended solids concentrations, coagulation usually occurs by enmeshment in insoluble hydrolysis products formed as the result of a reaction between the coagulant and the water. In this "sweep coagulation" the nature of the original suspended matter is of little significance and it is the properties of the hydrolysis product which control the reaction. Unfortunately, the behaviour of coagulants when added to water can be highly complex. In the case of aluminium sulphate the simplified reactions given earlier in this chapter are now known to be far removed from the actual situation. The hydrolysis products of aluminium are very complex, their nature being affected by such factors as the age and strength of the coagulant solution. Hydrolysis products of aluminium include compounds of the form

$$[Al(H_2O)_5OH]^{2+} \quad \text{and} \quad [Al_6(OH)_{15}]^{3+}$$

and sulphate complexes may also appear. As a result the actual reactions taking place are difficult to specify.

With higher suspended solids concentrations the colloidal theory can provide a basis for explaining the observed reactions. Thus destabilization of a colloidal suspension occurs due to the adsorption of strongly charged partially hydrolysed metallic ions. Continued adsorption results in charge reversal and restabilization of the suspension which does occur with high coagulant doses. In this type of situation the nature of the colloidal particles does therefore have an influence on the coagulation process.

When coagulation is used to remove colour from water the reaction appears to depend upon the formation of precipitates from the combi- nation of the soluble organics and the coagulant. There is thus generally a direct relationship between colour concentration and the dose of coagulant required for removal of the colour.

The value of coagulant aids can be related to the ability of large molecules in the form of long-chain structures to provide a bridging and binding action between adjacent suspended particles thus promoting agglomeration and preventing floc break-up under shear. With ionic coagulant aids, charge neutralization will also occur as with primary coagulants although at the normal doses employed for coagulant aids this effect is not likely to be very important.

Further Reading

ATKINSON, J. W., HILSON, M. A., BELL, F. and HUNTER, R. W., Practical experience in the use of polyelectrolytes. *Wat. Treat. Exam.* **20**, 1971, 165.

HOLLAND, G. J., Extensions to Hampton Load Works: Part II. Chemical aspects. *J. Instn Wat. Engrs*, **28**, 1974, 163.

IVES, K. J. and BHOLE, A. G., Theory of flocculation for continuous flow system. *J. Envr. Engng Div. Am. Soc. Civ. Engrs*, **99**, 1973, 17.

MCCOOKE, N. J. and WEST, J. R., The coagulation of a kaolinite suspension with aluminium sulphate. *Wat. Res.* **12**, 1978, 793.

PACKHAM, R. F. and SHEIHAM, I., Developments in the theory of coagulation and flocculation. *J. Instn Wat. Engrs Scits*, **31**, 1977, 96.

Problems

1. A flocculation tank 10 m long, 3 m wide and 3 m deep has a design flow of 0.05 m^3/s. Flocculation is achieved by three paddle wheels each with two blades 2.5 m by 0.3 m, the centre line of the blades being 1 m from the shaft which is at mid depth of the tank. The paddles rotate at 3 rev/min and the velocity of the water is 25 % of the blade velocity. For water at 20°C, $v = 1.011 \times 10^{-6}$ m^2/s. Calculate the power required for flocculation and the velocity gradient. (49.5 W, 23.2 m/s m)

2. A water supply with 15 mg/l alkalinity requires 40 mg/l aluminium sulphate for coagulation. Calculate the quantity of hydrated lime Ca(OH)$_2$ required to leave a finished water with 25 mg/l alkalinity. How much soda ash, Na$_2$CO$_3$, would be needed if it were used in place of lime? Ca 40, O 16, H 1, Na 23. (22.2 g/m^3, 31.8 g/m^3)

CHAPTER 13

Flow Through Porous Media

FILTRATION of suspensions through porous media, usually sand, is an important stage of the treatment of potable waters to achieve final clarification. Although about 90 % of the turbidity and colour are removed in coagulation and sedimentation a certain amount of floc is carried over from settling tanks and requires removal. Sand filtration is also employed to provide tertiary treatment of 30:20 standard sewage effluents. Other uses of flow through porous media include ion-exchange beds and absorption columns where the aim is not to remove suspended matter but to provide contact between two systems.

13.1. Hydraulics of Filtration

The resistance to flow of liquids through a porous medium is analogous to flow through small pipes and to the resistance offered by a fluid to settling particles.

The basic formulae for the hydraulics of filtration assume a bed of uni-size medium and refer to the schematic filter shown in Fig. 13.1.

The earliest formula due to Darcy is

$$\frac{h}{l} = \frac{v}{k} \tag{13.1}$$

where h = loss of head in bed of depth l with face velocity v,
k = coefficient of permeability.

Rose[1] used dimensional analysis to develop the equation

$$\frac{h}{l} = 1.067 \; C_D \frac{v^2}{gd\psi} \cdot \frac{1}{f^4} \tag{13.2}$$

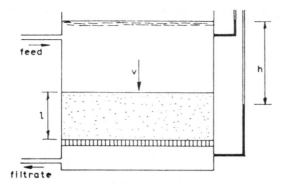

Fig. 13.1. Schematic filter.

where f = bed porosity = $\dfrac{\text{volume of voids}}{\text{total volume}}$,

d = characteristic diameter of bed particle,

ψ = particle shape factor,

C_D = Newton's drag coefficient = $\dfrac{24}{R} + \dfrac{3}{\sqrt{R}} + 0.34$.

The Carman–Kozeny[2] equation produces similar results

$$\frac{h}{l} = E \frac{(1-f)}{f^3} \frac{v^2}{gd\psi} \tag{13.3}$$

where $E = 150 \left(\dfrac{1-f}{R} \right) + 1.75.$

The particle shape factor ψ in equations (13.2) and (13.3) is the ratio of the surface area of the equivalent volume sphere to the actual surface area of the particle, i.e.

$$\psi = \frac{A_0}{A} \tag{13.4}$$

where A_0 = surface area of sphere of volume V.

For spherical particles ψ is unity and the particle diameter $d = (6V/A)$. For other shapes $d = (6V/\psi A)$.

Thus equations (13.2) and (13.3) can be rewritten as

$$\frac{h}{l} = 0.178 \, C_D \frac{v^2}{gf^4} \cdot \frac{A}{V} \tag{13.5}$$

$$\frac{h}{l} = E \left(\frac{1-f}{f^3} \right) \frac{v^2}{g} \cdot \frac{A}{6V} \tag{13.6}$$

Typical values of ψ are given in Table 13.1.

TABLE 13.1. TYPICAL VALUES OF
PARTICLE SHAPE FACTOR

Material	ψ
Mica flakes	0.28
Crushed glass	0.65
Angular sand	0.73
Worn sand	0.89
Spherical sand	1.00

Filters are normally used with graded sand, e.g. 0.5–1.0 mm, so that it is necessary to obtain an average A/V value:

$$\left(\frac{A}{V} \right)_{av} = \frac{6}{\psi} \sum \frac{p}{d} \tag{13.7}$$

where p = proportion of particles of size d (from sieve analysis). The slow sand filter with low hydraulic loading (about 2 m³/m² d) is cleaned by removal of the clogged surface layers and the bed is a homogeneous packing. In the case of the commoner rapid filter (about 120 m³/m² d) cleaning is by backwashing with filtrate from below thus producing a stratified bed packing and it is necessary to take account of the variation of C_D with particle size. Thus for a rapid filter using Rose's equation

$$\frac{h}{l} = 1.067 \frac{v^2}{g\psi f^4} \sum C_D \frac{p}{d} \tag{13.8}$$

Examples of this method of calculation are set out in the literature.[3, 4]

13.2. Filter Clogging

The equations above give the head loss with a clean bed, but when used for removal of suspended matter the porosity of the bed is continually changing due to the collection of particles in the voids. It is usually assumed that the rate of removal of particles is proportional to their concentration, i.e.

$$\frac{\partial c}{\partial l} = -\lambda c \tag{13.9}$$

where c = concentration of suspended solids,
  l = depth from inlet surface,
  λ = constant characteristic of the bed.

The equation is a partial differential because the solids concentration varies with time as well as position in the bed. Initially equation (13.9) may be integrated to give

$$\frac{c}{c_0} = e^{-\lambda_0 l} \tag{13.10}$$

where c_0 = solids concentration at inlet surface ($l = 0$).

Work by Ives and Gregory[5] has shown that if the bed retains solids the total head loss can be expressed as a linear function which for uni-size media is:

$$H = h + \frac{Kvc_0 t}{(1 - f)} \tag{13.11}$$

where h = head loss from the Carman–Kozeny equation,
  t = time of operation,
  K = constant.

For graded media the expression is similar, but with K replaced by another constant depending on the size grading and the variation of K with particle size.

Contrary to common belief the removal of suspended matter in a porous media bed is not simply a straining action. Removal of solids depends upon transport mechanisms such as gravity, interception, diffusion, sedimentation and hydrodynamic processes. Once transported into the pores of a bed the suspended matter is held there by attachment mechanisms due to physico-chemical and intermolecular forces similar to those which operate in coagulation. A bed of porous media is thus able to remove particles considerably smaller than the voids in the bed.

13.3. Filter Washing

With a slow filter, penetration of solids is superficial and cleaning is achieved by removing the upper layer of the medium at intervals of a few months, washing and replacing. The rapid filter clogs much more rapidly due to its higher hydraulic loading and the solids penetrate deep into the bed. Cleaning is achieved by backwashing at a rate of about ten times the normal filtration rate. The upward flow of water expands the bed producing a fluidized condition in which accumulated debris is scoured off the particles. Compressed air scouring prior to or at the same time as backwashing improves cleaning and reduces washwater consumption. Figure 13.2 illustrates the behaviour of a porous bed under backwash. As

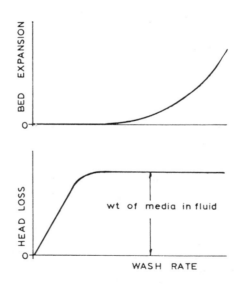

FIG. 13.2. Behaviour of filter bed during backwashing.

the backwash water is admitted to the bottom of the filter the bed begins to expand and there is an initial head loss. As the bed expands further the rate of increase of head loss decreases and when the whole bed is just suspended the head loss becomes constant. At this point the upward backwash force is equivalent to the downward gravitational force of the bed particles in water. Further increase in backwash flow increases the expansion but not the head loss. Excessive expansion is not desirable since the particles will be forced further apart, scouring action will be reduced and the backwash water consumption will be increased.

Referring to Fig. 13.3 showing a bed under backwashing conditions, the expansion is $(l_e - l)/l$; this is usually 5–25 % although expansions up to 50 % are sometimes used in the USA. The washwater velocity is often termed the rise rate.

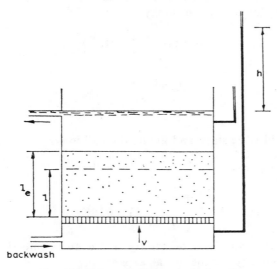

FIG. 13.3. Schematic filter under backwash.

At the maximum frictional resistance by the bed, upward water force

$$= \text{expanded depth} \times \text{nett unit weight of medium} \times \frac{\text{medium volume}}{\text{total volume}},$$

i.e.
$$hg\rho_w = l_e(\rho_s - \rho_w)g(1 - f_e) \qquad (13.12)$$

$$\therefore \qquad \frac{h}{l_e} = \frac{(\rho_s - \rho_w)}{\rho_w}(1 - f_e) \qquad (13.13)$$

or
$$\frac{h}{l_e} = (S_s - 1)(1 - f_e) \qquad (13.14)$$

where S_s = specific gravity of medium.

The particles are kept in suspension because of the drag force exerted on them by the rising water. Thus from settling theory equation (11.7)

$$C_D A \rho_w \frac{v^2}{2} \phi(f_e) = (\rho_s - \rho_w) g V \qquad (13.15)$$

$\phi(f_e)$ is introduced because v is the face velocity of the backwash water whereas the drag is governed by the particle settling velocity v_s.

It has been found experimentally[4] that

$$\phi(f_e) = \left(\frac{v_s}{v}\right)^2 = \left(\frac{1}{f_e}\right)^9. \qquad (13.16)$$

Thus
$$f_e = \left(\frac{v}{v_s}\right)^{0.22} \qquad (13.17)$$

or
$$v = v_s f_e^{4.5}$$

Now consider the static and fluidized conditions

$$(1 - f)l = (1 - f_e)l_e \qquad (13.18)$$

$$\frac{l_e}{l} = \frac{1 - f}{1 - f_e} = \frac{1 - f}{1 - \left(\dfrac{v}{v_s}\right)^{0.22}} \qquad (13.19)$$

For graded media it is necessary to use an arithmetic integration procedure to determine the overall expansion.

13.4. Types of Filter

Two basic types of filter have been used in the water industry for many years and their main characteristics are summarized in Table 13.2. The slow sand filter (Fig. 13.4) was the first type to be used and although some

TABLE 13.2. FILTER CHARACTERISTICS

Characteristic	Slow filter	Rapid filter
Filtration rate m³/m² d	1–4	100–200
Depth of bed m	1 (unstratified)	0.5–0.8 (stratified)
Size of bed m²	about 2000	about 100
Sand size mm	0.5–1.0	0.5–1.5
Max head loss m	1	2.5
Length of run d	20–90	1–5

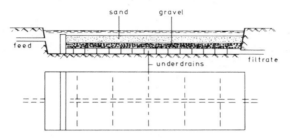

FIG. 13.4. Typical slow sand filter.

authorities consider the slow filter to be obsolete it does in fact still have many applications and may be particularly suitable in developing countries. Because of the low hydraulic loading there is only superficial penetration of suspended matter into the bed and runs of several weeks or months can be achieved with low turbidity inputs. Because of the long run times considerable biological activity occurs in a biological slime, the schmutz-decke, which forms on the surface of the bed. This slime layer contributes to the removal of fine suspended matter and often provides oxidation of organic contaminants in the raw water, which might otherwise cause taste and odour problems. Because of their large size the cleaning of slow filters, which usually involves removing and washing the top few centimetres of the bed, is costly. It is therefore important that slow filters are not used for raw waters regularly having more than about 20 units of turbidity. They are thus not suitable for use after chemical coagulation from which there is inevitably some carry-over of floc. With higher turbidity sources it is possible to utilize a double filtration process in which rapid filters remove most of the turbidity, the final removal and the oxidation of organic matter being achieved by secondary slow filters.

Rapid filters, which are widely used in water treatment and tertiary effluent treatment, normally operate under gravity conditions but may be used as pressure filters in circumstances where it is not desired to have a free water surface and thus destroy head. Figure 13.5 shows the main features of a conventional rapid gravity filter.

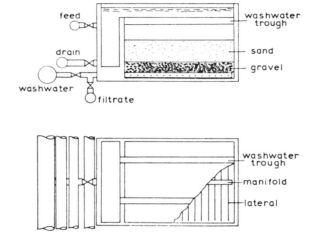

Fig. 13.5. Conventional rapid gravity filter.

The normal method of filtration downward through a bed of stratified medium with the finest particles at the top is clearly inefficient since the main solids load falls on the smallest pores. A more logical method would be to have the larger particles at the top thus reserving the smaller voids to trap the really fine particles. This situation can be achieved by upward filtration in which the feed water passes up through the bed which is back-washed in the normal way and thus the solids first meet the large bed particles. Care must be taken to control the filtration rate since high velocities will expand the bed and allow solids to escape. The use of a downflow filter comprising two media is also beneficial. A bed composed of a layer of anthracite (1.25–2.50 mm) which is less dense than the lower layer of 0.5 mm sand will remain in this configuration after backwashing and again has the advantage of presenting the large bed particles to the feed first. Such a filter will operate at a much lower head loss than a sand bed of the same overall depth at the same hydraulic loading and without deterioration in filtrate quality.

Filter runs may be terminated by one or more of the following criteria.

1. Terminal head loss. A clean rapid filter will have a head loss of about 0.3 m and the run may be terminated when the head reaches 2.5 m.
2. Duration of run. A simple time interval of, say, 24–72 h depending on feed quality may be satisfactory in conditions where the feed quality is relatively constant.
3. Filtrate quality. Continuous monitoring of filtrate turbidity may be used, the filter being backwashed as soon as the filtrate turbidity exceeds a predetermined value.

Whichever method is used, completely automatic operation of filters is possible.

The head loss across a filter bed increases during the run and most units incorporate a flow-control module which compensates for the increasing head loss in the bed so that the total head loss across the unit remains constant. It is, however, possible to operate filters on a declining rate basis with a constant head or obtain a constant output by allowing the inlet head to increase as the run proceeds (Fig. 13.6). These techniques are particularly

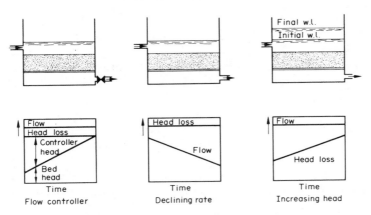

FIG. 13.6. Rapid filter-control techniques.

suitable for developing countries or small installations where the complexity and expense of flow controllers are undesirable.

References

1. ROSE, H. E., On the resistance coefficient–Reynolds number relationship for fluid flow through a bed of granular material. *Proc. Inst. Mech. Engrs*, **153**, 1945, 145.
2. CARMAN, P. C., Fluid flow through granular beds. *Trans. Inst. Chem. Engrs*, **15**, 1937, 150.
3. RICH, L. G., *Unit Operations of Sanitary Engineering*, John Wiley & Sons, New York, 1961, pp. 136–158.
4. FAIR, G. M., GEYER, J. C. and OKUN, D. A., *Waste Water Engineering: 2. Water Purification and Waste-Water Treatment and Disposal*, John Wiley & Sons, New York, 1967, 27.1–27.49.
5. IVES, K. J. and GREGORY, J., Basic concepts of filtration. *Proc. Soc. Wat. Treat. Exam.* **16**, 1967, 147.

Further Reading

ADIN, A., BAUMANN, R. E. and CLEASBY, J. L., The application of filtration theory to pilot-plant design. *J. Am. Wat. Wks Assn*, **71**, 1979, 17.
BOBY, W. M. T. and ALPE, G., Practical experiences using upward flow filtration. *Proc. Soc. Wat. Treat. Exam.* **16**, 1967, 215.
CLEASBY, J. L., STANGL, E. W. and RICE, G. A., Developments in backwashing of granular filters. *J. Envir. Engng Div. Am. Soc. Civ. Engrs*, **101**, 1975, 713.
GOULD, M. H. and PATTERSON, P., An experimental investigation into the cleaning of rapid sand filters. *Water Services*, **83**, 1979, 573.
IVES, K. J. (ed.), *The Scientific Basis of Filtration*, Noordhoff, Leyden, 1975.
IVES, K. J., A new concept of filtrability. *Prog. Wat. Tech.* **10**, 1978, 123.
LEEKAS, T. D., FOX, G. T. J. and MCNAUGHTON, J. G., Comparison of two and three layer granular bed filters for treating reservoir water. *J. Instn Wat. Engrs Scits*, **32**, 1978, 239.
TEBBUTT, T. H. Y., Filtration. In *Developments in Water Treatment*, **2** (Ed. Lewis, W. M.), Applied Science Publishers Ltd, Barking, 1980.

Problems

1. A rapid gravity filter installation is to treat a flow of 0.5 m^3/s at a filtration rate of 120 m^3/m^2 d with the proviso that the filtration rate with one filter washing is not to exceed 150 m^3/m^2 d. Determine the number of units and the area of each unit to satisfy these conditions. Each filter is washed for 5 min every 24 h at a wash rate of 10 mm/s, the filter being out of operation for a total of 30 min/d. Calculate the percentage of filter output used for washing.   (5, 72 m^3, 2.6%)

2. A laboratory-scale sand filter consists of a 10 mm diameter tube with a 900 mm deep bed of uniform 0.5 mm diameter spherical sand ($\psi = 1$), porosity 40%. Determine the head loss using Rose's formula and the Carman–Kozeny formula when filtering at a rate of 140 m³/m² d. For water at 20°C, $\mu = 1.01 \times 10^{-3}$ N s/m². (660 mm, 503 mm)

3. The measured settling velocity of the sand particles in the filter in question 2 was 100 mm/s. Determine the bed expansion when the filter is washed at a rate of 10 mm/s. (51%)

CHAPTER 14

Aerobic Biological Oxidation

THE amount of organic matter which can be assimilated by a stream is limited by the availability of DO as discussed in Chapter 7. In industrialized areas where large volumes of wastewater are discharged to relatively small rivers natural self-purification cannot maintain aerobic conditions and waste treatment additional to SS removal by physical means is essential. Removal of soluble and colloidal organic matter can be achieved by the same reactions as occur in self-purification, but more efficient removal can be achieved in a treatment plant by providing optimum conditions.

14.1. Principles of Biological Oxidation

The speed of an aerobic oxidation reaction cannot be greatly altered but by providing a large population of micro-organisms in the form of a slime or sludge it is possible to achieve a rapid rate of removal of organic matter from solution. The large microbial surface permits initial adsorption of colloidal and soluble organics together with synthesis of new cells so that after a relatively short contact time the liquid phase contains little residual organic matter. The adsorbed organic matter is then oxidized to the normal aerobic end products (Fig. 14.1).

The rate of removal of organic matter depends on the phase of the biological growth curve (Fig. 6.2). Figure 14.2 shows the growth curve with the aeration times employed for various forms of aerobic treatment. BOD theory assumes a first-order reaction and although some reactions are thought to be first order there is evidence, sometimes conflicting, that other reactions are zero order (i.e. independent of concentration) or second order. The situation becomes more complex when dealing with wastes such as sewage which contain many different compounds.

146

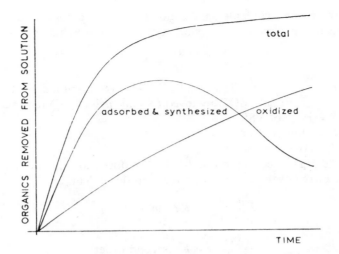

FIG. 14.1. Removal of soluble organics in biological treatment.

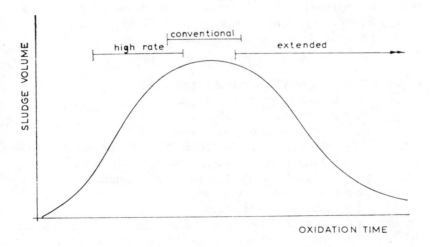

FIG. 14.2. Operating zones for biological oxidation processes.

At high organic contents the reaction is likely to be of zero order with constant rate of removal of organics per unit cell weight, i.e.

$$\frac{a}{S} \cdot \frac{dL}{dt} = K \tag{14.1}$$

where a = mass of volatile suspended solids (VSS) synthesized/unit mass ultimate BOD removed (or/unit mass COD removed),
S = mass of VSS,
L = ultimate BOD,
K = constant.

When the organic concentration has been reduced to some limiting value the rate of removal becomes concentration dependent, i.e.

$$\frac{a}{S} \cdot \frac{dL}{dt} = KL \text{ (first order)} \tag{14.2}$$

or

$$\frac{a}{S} \cdot \frac{dL}{dt} = KL^2 \text{ (second order)} \tag{14.3}$$

About one-third of the COD of a waste is used for energy and the remaining two-thirds is utilized for synthesis of new cells. Thus sludge production measured as VSS ranges from 0.2–0.8 kg/kg COD removed depending on the substrate and time of aeration. Allowance must be made for any SS initially present in the waste.

The volatile solids accumulation is given by

$$\text{VSA} = (aL_r + cS_i - S_e)q - bS_m qt \tag{14.4}$$

where VSA = mass of VSS accumulated/unit time,
S_i = concentration of VSS in influent,
S_e = concentration of VSS in effluent,
S_m = concentration of VSS in system,
L_r = concentration of ultimate BOD removed/unit time,
c = fraction of non-biodegradable VSS in influent,
b = endogenous respiration constant/unit time,
q = rate of flow/unit time,
t = retention time of system.

For aerobic oxidation a typical value of a would be 0.55 on an ultimate BOD basis (or $0.55 \times L/\text{BOD}_t$ for other than ultimate BOD values). A commonly used value of b is 0.15/day.

To maintain aerobic conditions oxygen must be supplied to the system and the oxygen requirement per unit time for the biochemical reactions may be obtained from

$$OR = 0.39L_r q + 1.42bS_m qt \tag{14.5}$$

14.2. Types of Aerobic Oxidation Plant

There are three basic types of aerobic unit:
(a) Percolating filter or bacteria bed.
(b) Activated sludge.
(c) Oxidation pond.
The biological filter and the activated-sludge process rely on similar principles, the oxidation pond which is only suited to warm climates operates somewhat differently utilizing bacteria and algae.

14.3. Percolating Filter

The oldest form of biological treatment unit consists basically of a bed of stone, circular or rectangular in plan (Fig. 14.3) with intermittent or continuous addition of settled sewage to the surface. On a conventional filter the medium is 50–100 mm grading, preferably a hard angular stone dosed by a rotating distributor mechanism, the normal depth of bed being 1.8 m.

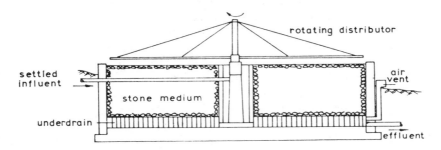

FIG. 14.3. A conventional percolating filter.

Liquid trickles through the interstices in the medium where micro-organisms grow in the protected areas forming a slime or film, the liquid flowing over the film rather than through it (Fig. 14.4). The micro-organisms are attracted to the medium by van der Waals' forces which are opposed by the shearing action of the liquid. Thus although there is little organic matter in solution in filter effluent there may be fairly high concentrations of SS in the form of displaced film and the effluent requires sedimentation in a humus tank to produce the desired effluent quality. The highest rate of oxidation takes place in the top section of the bed where the limiting factor is usually the amount of oxygen which can be supplied by natural ventilation (Fig. 14.5). Below this level the rate of oxidation decreases due to the decreasing concentration of organic matter in the liquid phase, and there is normally little benefit in using depths of medium greater than 2 m. The liquid film may only be in contact with the micro-organisms for a matter of 20–30 s, but because of the large surface area

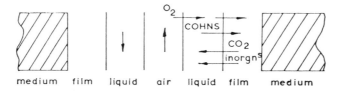

FIG. 14.4. Idealized section of a biological filter.

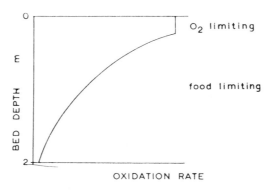

FIG. 14.5. Relation between depth of filter and rate of oxidation for a conventional unit.

available this contact time is sufficient for adsorption and stabilization. The maximum rate of stabilization occurs at the micro-organism/liquid interface since diffusion of organics through the film is slow. With a thick biological film, waste stabilization is not very efficient since much of the film is undergoing endogenous respiration.

Long experience in the UK has shown that to produce a 30:20 standard effluent after settlement, filters treating domestic sewage should be loaded at 0.07–0.10 kg BOD/m^2 d with a hydraulic loading of 0.12–0.6 m^3/m^3 d. If the filter loading is increased the higher content of organic matter will promote heavy film growths which may result in blockage of the voids causing ponding of the filter and anaerobic conditions. At conventional design loadings it is usual to obtain a fairly high degree of nitrification in the effluent although this will not be so apparent at the higher ranges of loading. Inevitably there have been many attempts during the years to produce more efficient percolating filters which would operate at much higher loadings and several modifications of the basic process have been produced (Fig. 14.6).

High-rate filtration. As mentioned previously excessive loading on filters results in ponding; this can be obviated or reduced by using large filter media and hydraulic loadings of 1.8 m^3/m^3 d will give 30:20 standard effluent from domestic sewage albeit with little or no nitrification. If a less stabilized effluent is required, e.g. roughing treatment for strong industrial wastes, loadings of 12 m^3/m^3 d with organic loadings of up to 1.8 kg BOD/m^3 d will give 60–70 % BOD removals. Treatment at such rates is facilitated by the use of plastic medium (90 % voids) in tall towers rather than the usual stone medium (40 % voids) the risk of ponding being thereby much reduced.

Alternating double filtration. A ponded filter can be brought back into use by applying the partially stabilized effluent from another filter. The film in ADF alternately grows and disintegrates, the total amount of film being less than in a single filter so that higher rates of loading can be safely employed. Two filters are operated in series and when the first filter shows signs of ponding the direction of flow through the filters is reversed. A second humus tank and additional pipework and pumping facilities are needed to operate ADF. A 30:20 standard effluent can be produced at loadings of 1.5 m^3/m^3 d and 0.24 kg BOD/m^3 d and the process is often useful in the relief of overloaded conventional percolating filters.

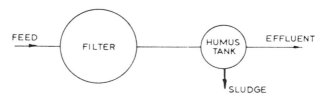

CONVENTIONAL

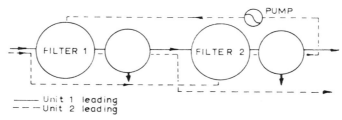

—— Unit 1 leading
— — —Unit 2 leading

ALTERNATING DOUBLE FILTRATION

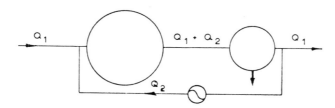

RECIRCULATION

FIG. 14.6. Modifications of the filtration process.

Recirculation. This is another method for increasing filter capacity based on the principle of treating settled waste in admixture with settled filter effluent in ratios of 1:0.5–10 depending on the strength of the waste. The concentration of organic matter in the feed to the filter is thus reduced at the expense of larger hydraulic loadings and additional pumping and pipework. Recirculation may be used only at low flows, at a constant rate or at rates proportional to the incoming flow. With particularly strong wastes two-stage recirculation may be adopted.

Nitrifying filters. To provide further stabilization and nitrification for an activated-sludge plant effluent high-rate filters are useful. At loadings of 9 m^3/m^3 d and 0.05–0.47 kg BOD/m^3 d a nitrified 30:20 standard effluent can be produced. Nitrifying bacteria also find application in water treatment plants dealing with raw waters containing significant amounts of ammonia. The nitrifying organisms may be contained in a high-rate bacteria bed or in a vertical-flow floc-blanket settling tank. In either case there may be an added benefit in that the micro-organisms can also oxidize trace organics present in the water which might otherwise cause taste and odour problems in the finished water.

Disc filters. The use of slowly rotating circular discs partially immersed in sewage provides a form of biological filter which is compact and suitable for off-site fabrication. Factory built units have a single tank divided by baffles into a number of chambers in which the disc surfaces are intermittently immersed as they rotate. Biological film grows on the discs and carries out oxidation of the sewage. Solids settle to the bottom of the tank which requires periodical desludging.

14.4. Activated Sludge

This process depends on the use of a high concentration of micro-organisms present as a floc kept suspended by agitation, originally with compressed air although mechanical agitation is also used now (Fig. 14.7). In either case high rates of oxygen transfer are possible. The effluent from the aeration stage is again low in dissolved organics but contains high SS (2000–8000 mg/l) which must be removed by sedimentation. The effectiveness of the process depends on the return of a portion of the separated sludge (living micro-organisms) to the aeration zone to recommence

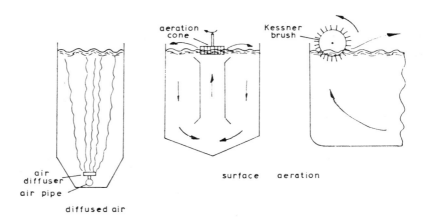

FIG. 14.7. Activated sludge aeration methods.

stabilization. The initial attractions of the activated-sludge process were that it occupied much less space than a percolating filter and had a much lower head loss. It has since proved useful for the treatment of many organic industrial wastes which were at one time thought to be toxic to biological systems.

In the diffused-air system much of the air is utilized for agitation and only a small amount is actually utilized for the oxidation reactions. In the absence of agitation (as in the final settling tank) the solids quickly settle to the bottom and lose contact with the organic matter in the liquid stage, the settled solids rapidly become anaerobic if not returned to the aeration zone. Sufficient air must be transferred to the mixed liquor to maintain a DO of 1–2 mg/l. The mixed liquor must be of suitable concentration and activity to give rapid adsorption and oxidation of the waste as well as providing a rapidly settling sludge so that a clarified effluent is produced quickly and the sludge can be returned to the aeration zone without delay.

In general a sludge volume of 25–50 % of the flow through the plant is drawn off from the settling tank and between 50 and 90 % of this is returned to the aeration zone, the remainder being dewatered and disposed of along with other sludges from the plant. If insufficient sludge is returned the mixed liquor suspended solids (MLSS) will be low and poor stabilization will result, the return of excessive amounts of sludge will result in very high

MLSS which may not settle well and which may exert higher oxygen demands than can be satisfied. If sludge is not removed rapidly from the settling tanks rising sludge may be found due to the production of nitrogen by reduction of nitrates under anaerobic conditions—a very poor effluent is then produced.

Because of the importance of maintaining good-quality sludge in the process various indices have been developed to assist in control.

1. *Sludge Volume Index SVI*

$$\text{SVI} = \frac{\text{settled volume of sludge in 30 min } (\%)}{\text{MLSS } (\%)} \qquad (14.6)$$

The SVI varies from about 40–100 for a good sludge, but may exceed 200 for a poor sludge with tendency to bulking. Bulking is used to describe a sludge with poor settling characteristics, often due to the presence of filamentous micro-organisms which tend to occur in plants with an easily degradable wastewater low in nitrogen and where the DO in the mixed liquor is low. A problem which arises when using SVI as a measure of the settling properties of a mixed liquor is that the values obtained are affected by the solids concentration and the diameter of the vessel used for the test. To overcome these problems the SSV (stirred specific volume) test is carried out at a fixed fluid SS concentration of 3500 mg/l and in standard 100 mm diameter vessel stirred at 1 rev/min.

2. *Sludge Density Index SDI*

$$\text{SDI} = \frac{\text{MLSS } (\%) \times 100}{\text{settled volume of sludge in 30 min } (\%)} \qquad (14.7)$$

SDI varies from about 2 for a good sludge to about 0.3 for a poor sludge.

3. *Mean Cell Residence Time θ_c*

$$\theta_c(\text{d}) = \frac{\text{aeration zone volume } (\text{m}^3) \times \text{MLVSS (mg/l)}}{\text{sludge wastage rate } (\text{m}^3/\text{d}) \times \text{sludge VSS (mg/l)}} \qquad (14.8)$$

For an activated-sludge plant producing a 30:20 standard effluent from normal settled sewage conventional design criteria are 0.56 kg BOD/m^3d with a nominal retention time in the aeration zone of 4–8 h. Nominal air supply is about 6.2 m^3/m^3, retention in the final settling tank is usually about 2 h. Such a plant will reliably produce a 30:20 standard effluent although nitrification may not be complete. A degree of nitrogen removal can be achieved by mixing settled sewage and return sludge in an anaerobic tank ahead of the conventional aeration tank.

Many modifications of the activated-sludge process have been produced both in the form of aeration (fine and coarse bubble diffusers, Inka system, high efficiency aerators with or without sparger rings) and in the actual process (Fig. 14.8).

High-rate activated sludge. With short retention times (2 h) and low MLSS (about 1000 mg/l) partial stabilization is achieved rapidly at low cost, at loadings of up to 1.6 kg BOD/m^3d with an air supply of about 3 m^3/m^3. Such plants achieve BOD removals of 60–70% and are suitable for pretreatment of strong wastes or for effluents discharged to estuarine waters where relaxed standards are applied.

Tapered aeration and step aeration. In a conventional flow-through system the rate of oxidation is highest at the inlet end of the tank and it may sometimes be difficult to maintain aerobic conditions there if a uniform air distribution is used. With tapered aeration the air supply is progressively reduced along the length of the tank so that although the same total volume of air is used as before more of the air is concentrated at the tank inlet to cope with the high demand there. Step aeration aims to achieve the same object by adding the waste feed at intervals along the tank to give a more constant oxygen demand in the aeration zone. It is not then necessary to reaerate the return sludge before addition to the aeration zone which is something often necessary in conventional plants to prevent zero DO at the aeration basin inlet zone. The completely mixed unit is the natural extension of this concept.

Extended aeration. Using long aeration times (24–48 h) it is possible to operate in the endogenous respiration zone so that less sludge is produced than in a normal plant. A low organic loading is used, 0.24–0.32 kg BOD/m^3d, and the plants have proved popular for small communities where the reduced sludge volume and the relatively inoffensive nature of the mineralized sludge are considerable benefits. These benefits are,

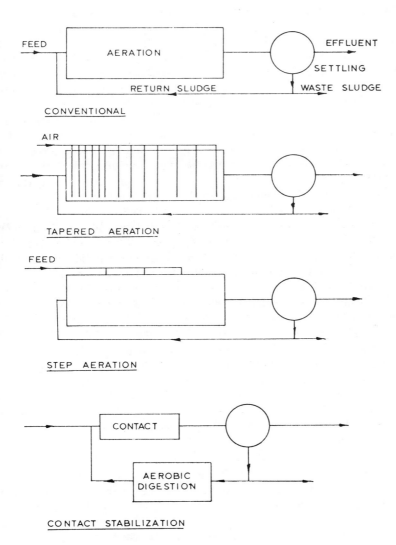

FIG. 14.8. Modifications of the activated-sludge process.

however, paid for in high operating costs (due to the long aeration time) and the plants do not normally produce a 30 : 20 standard effluent due to carry-over of solids from the settling zone.

Oxidation ditches. A development of the extended aeration process which has become popular of late is the adoption of brush-type aerators to provide aeration and to create motion in continuous ditches which can be relatively cheap to construct in suitable ground conditions.

Contact stabilization. This process uses the adsorptive capacity of the sludge to remove organic matter from solution in a small tank ($\frac{1}{2}$–1 h retention), the sludge and adsorbed organics then being transferred as a concentrated suspension to an aerobic digestion unit for stabilization (2–3 h retention). Solids in the contact zone are about 2000 mg/l, whereas in the digestion unit they may be as high as 20 000 mg/l.

Pure oxygen activated sludge. A recent development has been the introduction of activated-sludge plants operated with pure oxygen. Such installations involve the introduction of oxygen into closed stirred reaction tanks. It is claimed that these units can operate at relatively high MLSS levels (6000–8000 mg/l) whilst providing good sludge-settling charac-teristics and giving economies in power consumption and land area requirements. Operational data from pilot plants in the UK have not always confirmed the original claims, however.

Fluidized beds. In attempts to increase the population of micro-organisms in a biological treatment plant, and hence the efficiency of the suitable climatic conditions. Given sufficient land area they can give a very or plastics particles are being developed and show considerable promise.

14.5. Oxidation Pond

Oxidation ponds are shallow constructions, usually receiving raw sewage, which provide treatment by natural stabilization processes in suit-able climatic conditions. Given sufficient land area they can give a very satisfactory form of wastewater treatment in warm sunny climates. They are cheap to construct, simple to operate and provide good removals of organic matter and pathogenic micro-organisms. In some cases ponds may operate without producing an effluent due to evaporation and seepage but in most cases they are designed as continuous-flow systems.

Four main types of pond are used.

Faculative ponds. These are by far the most common and, as the name implies, combine aerobic and anaerobic activity in the same unit. Chlorophyll-bearing micro-organisms, phytoflagellates and algae operate in these ponds by utilizing the inorganic salts and carbon dioxide provided by the bacterial decomposition of organic matter as shown in Fig. 14.9. The oxygen produced by photosynthesis, which may give DO levels of 15–30 mg/l in late afternoon, is available for aerobic bacteriological activity although the DO level will fall during the night and may reach zero if the pond is overloaded. In the bottom deposits, anaerobic activity produces some stabilization of the sludge and releases some of the organic matter in soluble form for further degradation in the aerobic zone. Facultative ponds are usually 1–2 m deep with a surface loading of 0.02–0.05 kg BOD/m^2 d and nominal retention times of 5–30 d although these values can be altered by extreme temperatures. Because of the relatively long retention times and low organic concentration in such ponds there is a considerable removal of bacteria by endogenous respiration and by settlement. Bacteria and phytoplankton are preyed upon by ciliates, rotifers and crustaceans but some will escape in the effluent. Large algal growths occur and their presence in the effluent will produce moderate to high SS levels unless some means of harvesting or removal is employed. BOD removals of 70–85 % are possible although algae in the effluent can significantly affect these values. The most satisfactory shape for ponds is rectangular with a length:breadth of about 3:1. Simple earth banks are usually sufficient although in large

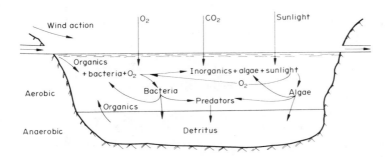

FIG. 14.9. Reactions in a facultative oxidation pond.

ponds, wave action may make bank protection with paving slabs, or similar materials, desirable. Shallow areas at the edges should be avoided to discourage mosquitoes and grass/weed cutting together with occasional insecticide spraying may be necessary. Solids will accumulate in the pond at a rate of 0.1–0.3 m^3/person year so that desludging will only be required at relatively long intervals of several years.

Maturation ponds. These are shallow fully aerobic ponds with a very low organic loading ($<$ 0.01 kg BOD/m^2 d) used primarily as a secondary stage of treatment following a facultative pond or other biological treatment unit. Again, large algal growths occur but their most important feature is the high removal of pathogenic bacteria because of the unfavourable environment for such organisms in the pond.

Anaerobic ponds. These are operated with a fairly high organic loading of about 0.5 kg BOD/m^{2}d with a depth of 3–5 m to ensure anaerobic conditions. They are capable of giving 50–60% BOD removal with a retention time of around 30 days and may be suitable for pretreating strong organic wastes before addition to facultative ponds. Anaerobic ponds are likely to produce odours so that they should not be sited near populated areas.

Aerated ponds. Analogous to the activated-sludge extended-aeration process these ponds utilize floating aerators to maintain DO levels and to provide mixing. BOD loadings of around 0.2 kg/m^2 d are possible with retention times of a few days and good-quality effluents are produced. The process involves the maintenance of an essentially bacterial floc rather than the bacterial/algal system of the simpler ponds. The need for mechanical plant and reliable power supply detracts from the basic simplicity of the oxidation pond but aerated ponds can have applications in the urban areas of developing countries.

Further Reading

ARTHUR, J. P., The development of design equations for facultative waste stabilization ponds in semi-arid areas. *Proc. Instn Civ. Engrs*, **71**, 1981, Part 2, 197.

BAKER, J. M. and GRAVES, Q. B., Recent approaches for trickling filter design. *J. San. Engng Div. Am. Soc. Civ. Engrs*, **94**, 1968, 61.

BARNES, D., FORSTER, C. F. and JOHNSTONE, D. W. M. (ed.), *Oxidation Ditches in Wastewater Treatment*, Pitman Books Ltd., London, 1982.

BRUCE, A. M., BROWN, B. L. and MANN, H. T., Some developments in the treatment of sewage from small communities. *Pub. Hlth Engr*, 1973 (3), 116.
CHRISTOULAS, D. G. and TEBBUTT, T. H. Y., A simple model of the complete-mix activated-sludge process. *Envr. Tech. Ltrs*, **3**, 1982, 89.
COOPER, P. F. and WHEELDON, D. H. V., Fluidized and expanded-bed reactors for waste-water treatment. *Wat. Pollut. Control*, **79**, 1980, 286.
CURI, K. and ECKENFELDER, W. W. (eds.), *Theory and Practice of Biological Wastewater Treatment*, Sijthoff & Nordhoff, Alphen aan den Fijn, 1980.
ELLIS, K. V. and BANAGA, S. E. I., A study of rotating-disc treatment units operating at different temperatures. *Wat. Pollut. Control*, **75**, 1976, 73.
GLOYNA, E. F., *Waste Stabilization Ponds*. WHO, Geneva, 1971.
LUMBERS, J. P., Waste stabilization ponds: Design considerations and methods. *Pub. Hlth Engr*, **7**, 1979, 70.
TEBBUTT, T. H. Y. and CHRISTOULAS, D. G., Performance studies on a pilot-scale activated-sludge plant. *Wat. Pollut. Control*, **74**, 1975, 701.
TEBBUTT, T. H. Y. and AZIZ, J. A., An assessment of simple mathematical models for the activated-sludge process. *Envr. Tech. Ltrs*, **1**, 1980, 440.
WHITE, M. J. D., Design and control of secondary settlement tanks. *Wat. Pollut. Control*, **75**, 1976, 419.
WOOD, L. B., KING, R. P., DURKIN, M. K., FINCH, H. J. and SHELDON, D., The operation of a simple activated sludge plant in an atmosphere of pure oxygen. *Pub. Hlth Engr*, **4**, 1976, 36 and 67.

Problems

1. An activated-sludge plant with 3000 mg/l MLVSS treats a waste with an ultimate BOD of 1000 mg/l having 350 mg/l VSS which are 90% biodegradable. The plant effluent contains 30 mg/l ultimate BOD and 20 mg/l VSS. The hydraulic retention time of the system is 6 h. Determine the daily VSS accumulation and the oxygen requirement for a flow of 0.1 m^3/s if the synthesis constant (a) is 0.55 and the endogenous respiration constant (b) is 0.15 d^{-1}. (3767 kg/d, 4648 kg/d)

2. Control analyses on an activated-sludge plant indicated MLSS 4500 mg/l and solids settled in 30 min 25%. The plant treats a waste with 300 mg/l SS and a flow of 0.1 m^3/s. Aeration zone capacity is 2500 m^3. Sludge wastage rate 100 m^3/d with VSS 15 000 mg/l. Calculate the SVI, SDI, and Mean Cell Residence Time. (55.5, 1.8, 7.5 d).

3. Compare the area requirements for conventional filters (0.1 kg BOD/m^3 d) and conventional activated sludge (0.56 kg BOD m^3 d) for the flow from a town of 28 000 population, d.w.f. 200 l/person day with 250 mg/l BOD. Assume a filter depth of 2 m and 3 m deep aeration tanks. Primary sedimentation removes 35% of the applied BOD. (4550 m^2, 542 m^2)

4. A West African village has a population of 500 people and a daily sewage flow of 45 l/person. The *per capita* BOD contribution is 0.045 kg/d. Determine suitable dimensions for a facultative oxidation pond to serve the village if the appropriate loading is 0.050 kg BOD m^2 d and the desired retention time is 20 d. (36.75 × 12.25 × 1 m)

CHAPTER 15

Anaerobic Biological Oxidation

WITH very strong organic wastes containing high SS and with the sludges from primary sedimentation and aerobic treatment it becomes difficult to maintain aerobic conditions. The physical limitations of oxygen transfer equipment may prevent satisfaction of the oxygen demand with consequent onset of anaerobic conditions. In such circumstances it is possible to achieve partial stabilization by anaerobic oxidation or digestion.

15.1. Principles of Anaerobic Oxidation

Anaerobic oxidation obeys the same general laws as aerobic oxidation so that equations (14.1), (14.2), (14.3) and (14.4) may be applied. The methane produced by anaerobic oxidation is of some value as a fuel and the volume produced from a particular organic compound can be determined from the following relation:

$$C_nH_aO_b + \left(n - \frac{a}{4} - \frac{b}{2}\right)H_2O \rightarrow \left(\frac{n}{2} - \frac{a}{8} + \frac{b}{4}\right)CO_2 + \left(\frac{n}{2} + \frac{a}{8} - \frac{b}{4}\right)CH_4$$

(15.1)

At STP 1 kg ultimate BOD (or 1 kg COD) oxidized anaerobically yields about $0.35\,m^3$ methane gas which has a calorific value of $35\,kJ/l$.

The rate of gas production is temperature dependent as shown in Fig. 15.1. Optimum gas production occurs at $35°C$ (mesophilic digestion) and $55°C$ (thermophilic digestion). The higher temperature is normally only economic in warm climates because of the high heat loss in other regions. In practice when allowing for synthesis, methane production can be estimated from the formula

$$G = 0.35\,(L_r q - 1.42\,VSA)$$

(15.2)

162

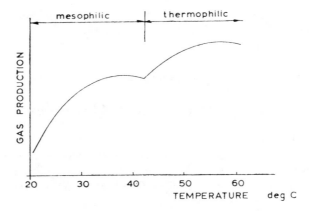

FIG. 15.1. Effect of temperature on gas production.

where G = volume of CH_4 produced m^3/unit time,

  L_r = concentration of ultimate BOD removed/unit time,

  q = rate of flow/unit time,

  VSA = mass of VSS accumulated/unit time.

The mass of VSS produced/unit time in an anaerobic process can be obtained from

$$\text{VSA} = \frac{aL_rq}{1 + bt_s} \qquad (15.3)$$

where a = mass of VSS synthesized/unit mass ultimate BOD removed,

  b = endogenous respiration constant/unit time,

  t_s = solids retention time = $\dfrac{\text{mass of solids in system}}{\text{solids accumulation/unit time}}$.

15.2. Applications of Digestion

The main use of anaerobic digestion is for the stabilization of primary and secondary sludges which have solids contents of 20 000–60 000 mg/l (2–6 %) about 70 % of which are volatile. The solids in primary sludge are readily putrescible and have a heavy faecal odour. The result of digestion is to reduce the volatile content to about 50 % and the total solids to about 70 % of the original values. The remaining organic solids are homogeneous in nature and are relatively stable, having a tarry odour. Dewatering of the digested sludge is difficult, however.

Conventional anaerobic digestion is carried out as a two-stage process (Fig. 15.2), the first stage heated to the desired temperature produces most of the gas and the second stage is one of quiescent settlement and solids separation. The supernatant liquor which is high in soluble organics (up to 10 000 mg/l BOD) is drawn off for aerobic treatment in the main plant, and the settled solids are withdrawn as a sludge for dewatering and ultimate disposal. Common design criteria are volatile solids loadings of 0.5–1.0 kg/m³ d with total retention times of 20–60 d.

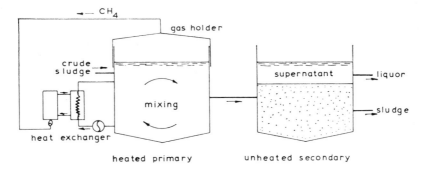

FIG. 15.2. Conventional sludge-digestion plant.

High-rate digestion which provides for solids separation and recycling similar to the activated-sludge process, permits retention times of 10–30 h at loadings of up to 2.0 kg VS/m³ d. The actual retention time of the solids is in the region of 10–20 d and a degasifier is usually required to minimize floating solids in the separation stage. High-rate processes have application for the treatment of strong organic wastes (1–5 % VS) where BOD removals of 90 % can be achieved. Because of the high initial strength the effluent normally requires further treatment before discharge to a watercourse but this can be obtained with a small aerobic unit.

An interesting development in anaerobic processes is the use of anaerobic filters where the biological growth is maintained on sand or plastic media and enables good contact to be provided between the organisms and the organic matter in the waste.

15.3. Operation of Digesters

As outlined in Chapter 6 the anaerobic process is sensitive to acid pH conditions and requires careful control.

For good digestion the pH is usually between 6.5 and 7.5 and a falling pH means that the process is becoming unbalanced. Excess production of volatile acids destroys the buffering capacity of alkalinity in the sludge, lowers the pH and reduces gas production (Fig. 15.3). As long as the sludge has a fairly high alkalinity an increase in acid production may initially produce little effect on pH so that measurement of volatile acids is a better control parameter. The normal volatile acid content is 250–1000 mg/l and if it exceeds 2000 mg/l trouble is likely. Lime is often used to aid recovery of digestion after high acid production.

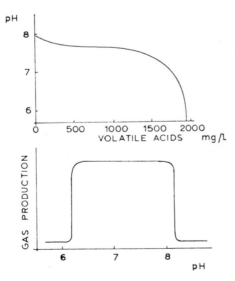

Fig. 15.3. Effect of volatile acids production on digestion.

The changes which occur in a simple batch-fed digester are shown in Fig. 15.4. The initial drop in pH occurs because of the faster action of the acid-forming bacteria. As the methane formers build up the acid content is reduced and gas production increases. The initial start-up of digestion is achieved by seeding with active sludge from another plant, or by starting with partial load (about one-tenth of the normal) and slowly increasing. These methods prevent excessive acid production from inhibiting growth of methane bacteria.

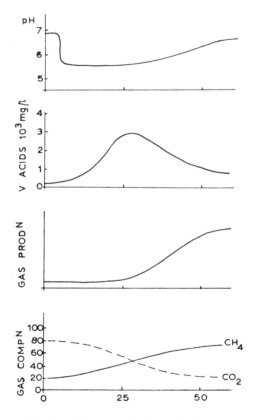

FIG. 15.4. Behaviour of a batch-fed digester.

In small treatment plants digestion is sometimes carried out in unheated tanks with no facilities for gas collection. Such a procedure is only satisfactory in warm climates since in temperate zones active digestion occurs only during the summer. The septic tank used for single houses and small communities is in fact an anaerobic oxidation plant which removes SS from sewage and breaks them down anaerobically. Septic tank effluent whilst low in SS still has a high BOD and should be treated on a biological filter before discharge to a watercourse.

Further Reading

ANDERSON, G. K. and DONNELLY, T., Anaerobic digestion of high strength industrial wastewaters. *Pub. Hlth Engr*, **5**, 1977, 64.

BRADE, C. E. and NOONE, G. P., Anaerobic sludge digestion—need it be expensive? Making more of existing resources. *Wat. Pollut. Control*, **80**, 1981, 70.

BRUCE, A. M., New approaches to anaerobic sludge digestion. *J. Instn Wat. Engrs Scits*, **35**, 1981, 215.

FROSTELL, B., Anaerobic treatment in a sludge bed system compared with a filter system. *J. Wat. Pollut. Control Fedn*, **53**, 1981, 216.

GRIFFITHS, J., The practice of sludge digestion. In *Waste Treatment* (Ed. Isaac, P. C. G.), Pergamon Press, Oxford, 1960, p. 367.

HOBSON, P. N., BOUSFIELD, S. and SUMMERS, R., *Methane Production from Agricultural and Domestic Wastes*, Applied Science Publishers Ltd., Barking, 1981.

SPEECE, R. E. and McCARTY, P. L., Nutrient requirements and biological solids accumulation in anaerobic digestion. In *Advances in Water Pollution Research. Proc. 1st Intl. Conf. Water Poll. Res. 2*, Pergamon Press, New York, 1964, p. 305.

Problems

1. An anaerobic digestion plant is to give 90% BOD removal to $100\,m^3/d$ of slaughterhouse effluent with an ultimate BOD of $3500\,mg/l$. The solids retention time is 10 d. Calculate the daily solids accumulation and the daily gas production. Synthesis constant (a) is 0.1 and the endogenous respiration constant (b) is $0.01\,d^{-1}$. ($28.6\,kg$, $96\,m^3$)

2. In a treatment plant $250\,m^3$ of primary sludge are produced daily with a total solids content of 5%, volatile matter is 65% of the total solids. Determine the anaerobic digester capacity required for a loading of $0.75\,kg\,VS/m^3$ d and calculate the nominal retention time. ($10\,830\,m^3$, 43.3 d)

CHAPTER 16

Disinfection

BECAUSE of the small size of many micro-organisms it is not possible to guarantee their complete removal from water by such forms of treatment as coagulation and filtration. Some form of disinfection is therefore necessary to ensure the elimination of potentially harmful micro-organisms from potable waters. Most wastewaters and treated effluents contain large numbers of micro-organisms and it may sometimes be appropriate to disinfect such liquids although in general the routine disinfection of effluents is not to be recommended. Disinfection of an effluent before discharge will tend to retard self-purification reactions in the receiving water and the formation of reaction products from the interactions of organic compounds and disinfectants may be undesirable in waters used as sources for potable supply.

It is important to appreciate the difference between sterilization (the killing of all organisms) which is rarely practised or needed, and disinfection (the killing of potentially harmful organisms) which is the normal requirement.

16.1. Theory of Disinfection

In general the rate of kill is given by

$$\frac{dN}{dt} = -KN \tag{16.1}$$

where K = reaction rate constant for particular disinfectant,
N = number of viable organisms.
Integrating

$$\log_e \frac{N_t}{N_0} = -Kt \tag{16.2}$$

168

where N_0 = number of organisms initially,
N_t = number of organisms at time t.
Changing to base 10

$$\log \frac{N_t}{N_0} = -kt \tag{16.3}$$

where $k = 0.4343\,K$

or

$$t = \frac{1}{k}\log \frac{N_0}{N_t} \tag{16.4}$$

Since N_t will never reach zero it is normal to specify kill as a percentage, e.g. 99.9 %. The rate constant as well as depending on the particular disinfectant also varies with disinfectant concentration, temperature, pH and other environmental factors.

The most popular disinfectant for water is chlorine which does not obey equation (16.1) but follows the relation

$$\frac{dN}{dt} = -KNt \tag{16.5}$$

or integrating and changing to base 10

$$t^2 = \frac{2}{k}\log \frac{N_0}{N_t} \tag{16.6}$$

at pH 7, values of k for chlorine are about 1.6×10^{-2}/s for free residuals and 1.6×10^{-5}/s for combined residuals, when applied to coliform organisms.

16.2. Chlorine

Chlorine (and its compounds) is widely used for the disinfection of water because:
1. It is readily available as gas, liquid, or powder.
2. It is cheap.
3. It is easy to apply due to relatively high solubility (7000 mg/l).
4. It leaves a residual in solution which while not harmful to man provides protection in the distribution system.
5. It is very toxic to most micro-organisms, stopping metabolic activities.

It has some disadvantages in that it is a poisonous gas which requires careful handling and it can give rise to taste and odour problems particularly in the presence of phenols.

Chlorine is a powerful oxidizing agent which will rapidly combine with reducing agents and unsaturated organic compounds, e.g.

$$H_2S + 4Cl_2 + 4H_2O \rightarrow H_2SO_4 + 8HCl$$

This immediate chlorine demand must be satisfied before chlorine becomes available for disinfection. 1 mg/l of chlorine will oxidize 2 mg/l BOD but this is not normally a feasible method of wastewater treatment.

After the chlorine demand has been satisfied the following reactions can occur:

1. In the absence of ammonia:

$$Cl_2 + H_2O \rightleftharpoons HCl \quad + \quad HClO$$
$$\updownarrow \qquad \qquad \updownarrow \qquad \text{(free residuals)}$$
$$H^+ + Cl^- \quad H^+ + ClO^-$$

Hypochlorous acid HClO is the more effective disinfectant, the chlorite ion ClO^- being relatively ineffective. The dissociation of HClO is suppressed at acid pH values, the residual being all HClO at pH 5 and below, about half HClO at pH 7.5 and all ClO^- at pH 9. Thus the most effective disinfection occurs at acid pH levels.

2. In the presence of ammonia:

$$NH_4^+ + HClO \rightarrow NH_2Cl + H_2O + H^+$$
$$\text{monochloramine}$$

$$NH_2Cl + HClO \rightarrow NHCl_2 + H_2O \qquad \text{(combined residuals)}$$
$$\text{dichloramine}$$

$$NHCl_2 + HClO \rightarrow NCl_3 + H_2O$$
$$\text{nitrogen trichloride}$$

The combined residuals are more stable than free residuals but less effective as disinfectants. For a given kill with constant residual the combined form requires a hundred times the contact time required by the free residual. Alternatively for a constant contact time the combined residual concentration must be twenty-five times the free residual concentration to give the desired kill.

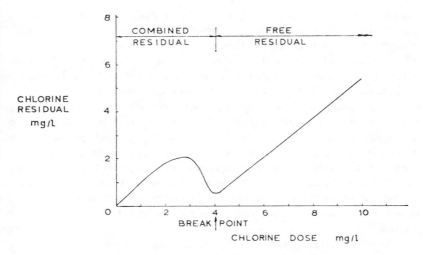

FIG. 16.1. Chlorine residuals for a water with 0.5 mg/l Amm.N.

Ammonia may be added to waters lacking in ammonia to allow the formation of chloramines which tend to cause less trouble with tastes and odours than do free residuals. In the presence of ammonia the continued addition of chlorine produces a characteristic residual curve as shown in Fig. 16.1. Once all the ammonia has reacted further chlorine converts the combined residual into a free residual at the breakpoint, simplified as

$$NCl_3 + Cl_2 + H_2O \rightarrow HClO + NH_4^+$$

The breakpoint theoretically occurs at 2 parts Cl_2:1 part NH_3 but in practice the ratio is nearer $10:1$. Past the breakpoint the free residual is proportional to the dose.

Troublesome tastes and odours can be destroyed using the oxidizing action of excess chlorine in the process known as superchlorination, the excess chlorine being removed by sulphonation after the desired contact time.

$$Cl_2 + SO_2 + H_2O \rightarrow H_2SO_4 + HCl$$

Considerable concern has recently been expressed about the presence in water of small concentrations of organochlorine compounds some of which are carcinogenic at relatively high doses in animals. There thus may be a potential hazard in the lifetime consumption of drinking water with concentrations of a few $\mu g/l$ of these compounds. The organochlorines, of which trihalomethanes such as chloroform are the most common, occur when raw waters containing organic matter, e.g. from sewage effluents, are disinfected with chlorine. There is no scientific evidence that the levels currently found in water supplies are in any way hazardous but it is sensible to endeavour to prevent their formation by careful process control and avoidance of unnecessary use of chlorine. Additional pressures are developing in some quarters against the use of gaseous chlorine because of the potential hazard arising from a gas leak at the treatment plant or during the transport of bulk chlorine. There has thus been a tendency to consider the use of hypochlorite solution for disinfection, particularly in small plants, where safety measures for chlorine gas may be more difficult to arrange.

Chlorine dioxide is mainly used for the control of tastes and odours since although there is some evidence that it is a more powerful disinfectant than chlorine in alkaline conditions, it is much more expensive. It does not combine with ammonia to any appreciable extent so that chlorine dioxide can be used to obtain free chlorine residuals in a water with large amounts of ammonia. The presence of significant amounts of ammonia in a water supply is, however, undesirable because of its nutrient value which tends to encourage biological growth. Chlorine dioxide is unstable and must be generated *in situ* by the action of chlorine or an acid on sodium chlorite

$$2NaClO_2 + Cl_2 \rightarrow 2ClO_2 + 2NaCl$$
$$5NaClO_2 + 4HCl \rightarrow 4ClO_2 + 5NaCl + 2H_2O$$

The possible formation of organochlorine compounds by chlorine dioxide has not yet been fully investigated so that its use in place of chlorine may not necessarily solve the trace organics problem.

16.3. Ozone

Ozone (O_3) is an allotropic form of oxygen produced by passing dry oxygen or air through an electrical discharge (5000–20 000 V, 50–500 Hz). It is an unstable, highly toxic blue gas with a pungent odour of new mown hay. A powerful oxidizing agent it is an efficient disinfectant and useful in bleaching colour and removing tastes and odours. Like oxygen it is only slightly soluble in water and because of its unstable form it leaves no residual. Unless cheap energy is available ozone treatment is much more expensive than chlorination but it does have the advantage of good colour removal. In these circumstances, filtration and ozonization may give a finished water similar to that produced by a more complex coagulation, sedimentation, filtration and chlorination plant. Because of the absence of ozone residuals in the distribution system, biological growths with attendant colour, taste and odour problems may result. Such growths in the distribution system can usually be prevented by adding a small dose of chlorine after ozonization. Ozone has some application in the oxidation of certain industrial wastewaters not amenable to biological oxidation.

Ozone must be manufactured on site by passing dry air through a high-voltage high-frequency electrical discharge. There are two main types of ozonizer; plate type with flat electrodes and glass dielectrics, and the tube type with cylindrical electrodes coaxial with glass dielectric cylinders. The high-tension side is cooled by convection and the low-tension side by water. Air passes between the electrodes and is ozonized by the discharge across the air gap. Ozone production is usually up to about 4% by weight of the carrier air with power requirements of around 25 kWh/kg of ozone produced. Ozone will react with organic matter to form ozonides in certain conditions and the significance of the presence of these products in water is not yet fully understood.

16.4. Ultraviolet Radiation

Various forms of radiation can be effective disinfecting agents and UV radiation has been used for the treatment of small water supplies for many years. The disinfecting action of UV at a wavelength of around 254 nm is quite strong provided that the organisms are actually exposed to the radiation. It is thus necessary to ensure that turbidity is absent and that the dose is increased to allow for the absorption of UV by any organic compounds present in the flow. The water to be disinfected flows between mercury arc discharge tubes and polished metal reflector tubes which gives efficient disinfection with a retention time of a few seconds although at a rather high power requirement of 10–20 W/m^3 h. The advantages of UV disinfection include: no formation of tastes and odours, minimum maintenance, easy automatic control with no danger from overdosing. Disadvantages are: lack of residual, high cost and need for high clarity in the water.

16.5. Other Disinfectants

Various other methods have been used for water disinfection:

1. *Heat.* Disinfection by heat is very effective but is costly and impairs the palatability of water by removing DO and dissolved salts. No residual effect.

2. *Silver.* Colloidal silver was used by the Romans to preserve the quality of water in storage jars since, at concentrations of about 0.05 mg/l, silver is toxic to most micro-organisms. It is of value for small portable filter units for field use where silver-impregnated gravel filter candles remove turbidity and provide disinfection. The cost becomes excessive for other than very small supplies.

3. *Bromine.* A halogen like chlorine, bromine has similar disinfection properties and is sometimes used in swimming pools where the residual tends to be less irritating to the eyes than chlorine residuals.

Further Reading

BRETT, R. W. and RIDGWAY, J. W., Experiences with chlorine dioxide in Southern Water Authority and Water Research Centre. *J. Instn Wat. Engrs Scits,* **35**, 1981, 135.

CAMPBELL, R. M. and PESCOD, M. B., The ozonization of Turret and other Scottish waters. *J. Instn Wat. Engrs,* **19**, 1965, 101.

JEPSON, J. D., Disinfection of water supplies by ultra-violet radiation. *Wat. Treat. Exam.,* **22**, 1973, 175.

RICHARDS, W. N. and SHAW, B., Developments in the microbiology and disinfection of water supplies. *J. Instn Wat. Engrs Scits,* **30**, 1976, 191.

Problems

1. Ozone is to be used to obtain a 99.9 % kill of bacteria in water with a residual of 0.5 mg/l. Under these conditions the reaction constant (k) is 2.5×10^{-2}/s. Determine the contact time required. (120 s)

2. Compare the contact times necessary to give *E. coli* kills of 99.99 % in water with (*a*) free chlorine residual of 0.2 mg/l, and (*b*) combined chlorine residual of 1 mg/l. k values are 10^{-2}/s and 10^{-5}/s respectively. (28 s, 890 s)

CHAPTER 17

Chemical Treatment

A NUMBER of constituents of waters and wastewaters do not respond to the conventional treatment processes already discussed, and alternative forms of treatment must be used for their removal. Soluble inorganic matter can often be removed by precipitation or ion-exchange techniques. Soluble non-biodegradable organic substances may frequently be removable by adsorption.

17.1. Chemical Precipitation

Removal of certain soluble inorganic materials can be achieved by the addition of suitable reagents to convert the soluble impurities into insoluble precipitates which can then be flocculated and removed by sedimentation. The extent of removal which can be accomplished depends on the solubility of the product; this is usually affected by such factors as pH and temperature.

Chemical precipitation may be used in industrial wastewater treatment, e.g. to remove toxic metals from metal-finishing effluents. Such effluents often contain considerable amounts of hexavalent chromium which is harmful to biological systems. By the addition of ferrous sulphate and lime the chromium is reduced to the trivalent form which can be precipitated as a hydroxide.

$$Cr^{6+} + 3Fe^{++} \rightarrow Cr^{3+} + 3Fe^{3+}$$
$$Cr^{3+} + 3OH^- \rightarrow \underline{Cr(OH)_3}$$
$$Fe^{3+} + 3OH^- \rightarrow \underline{Fe(OH)_3}$$

For efficient treatment it is essential to add the correct dose of reagents. For chromium reduction the theoretical requirement is as shown above, but at this level the reaction proceeds very slowly and in practice to ensure complete reduction it is necessary to add 5–6 atoms of ferrous iron for each atom of hexavalent chromium. A characteristic of chemical precipitation processes is the production of relatively large volumes of sludge.

The most widespread use of chemical precipitation is in water softening. Hard waters, i.e. waters containing calcium and magnesium in significant amounts, often require softening to improve their suitability for washing and heating purposes. For potable supplies water with up to about 75 mg/l hardness (as $CaCO_3$) is usually considered as soft, but some surfacewaters and many groundwaters may have hardness levels of several hundred mg/l. A hardness in excess of 300 mg/l would normally be considered undesirable. The need for softening of domestic supplies depends on reasons of convenience and economy rather than of health, since even at very high concentrations (> 1000 mg/l) hardness is quite harmless. Indeed there is some statistical evidence to suggest that artificially softened waters may increase the incidence of some forms of heart disease.

Hardness is normally expressed in terms of calcium carbonate whereas chemical analyses for individual ions are usually given in terms of that ion. It is thus necessary to convert the analytical results to the common denominator.

$$X \text{ mg/l as } CaCO_3 = X \text{ mg/l} \times \frac{\text{equiv wt } CaCO_3}{\text{equiv wt } X} \qquad (17.1)$$

where X is any ion or radical.

Thus for a typical water analysis conversion may be carried out as follows:

$$40 \text{ mg/l Ca}^{++} \times \frac{50}{20.04} = 99.0 \text{ mg/l as } CaCO_3$$

$$24 \text{ mg/l Mg}^{++} \times \frac{50}{12.16} = 98.5 \text{ mg/l as } CaCO_3$$

$$9.2 \text{ mg/l Na}^{+} \times \frac{50}{23} = 20.0 \text{ mg/l as } CaCO_3$$

$$183 \text{ mg/l HCO}_3^{-} \times \frac{50}{61} = 150.0 \text{ mg/l as } CaCO_3$$

$$57.5 \text{ mg/l } SO_4^{=} \times \frac{50}{48} = 58.0 \text{ mg/l as } CaCO_3$$

$$7.0 \text{ mg/l } Cl^{-} \times \frac{50}{35.5} = 9.5 \text{ mg/l as } CaCO_3$$

Note that the cations and anions both sum to 217.5 mg/l. It is now possible to represent the composition of the water as a bar diagram (Fig. 17.1).

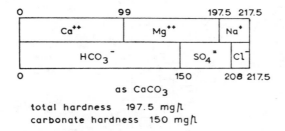

FIG. 17.1. Composition of a water sample in terms of $CaCO_3$.

Precipitation softening is based on the reversal of the process by which the hardness entered the water initially, i.e. the conversion of soluble compounds into insoluble ones which will then precipitate and permit removal by flocculation and sedimentation. The method of precipitation softening adopted depends on the form of hardness present.

1. *Lime softening.* For calcium hardness of the carbonate form. The addition of lime equivalent to the amount of bicarbonate present will form insoluble calcium carbonate

$$Ca(HCO_3)_2 + Ca(OH)_2 \rightarrow \underline{2CaCO_3} + 2H_2O$$

The solubility of $CaCO_3$ at normal temperatures is about 20 mg/l and because of the limited contact time available in a normal plant a residual of about 40 mg/l $CaCO_3$ usually results. The softened water is thus saturated with $CaCO_3$ and deposition of scale would be likely in the distribution system, this may be prevented by carbonation which produces soluble $Ca(HCO_3)_2$

$$CaCO_3 + CO_2 + H_2O \rightarrow Ca(HCO_3)_2$$

or by the addition of phosphates which sequester calcium and prevent scaling. Figure 17.2 shows the steps in lime softening.

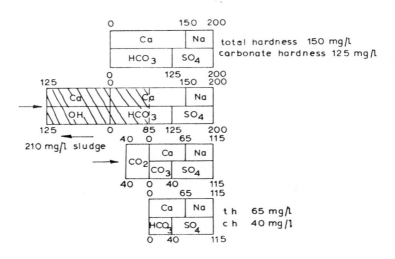

FIG. 17.2. Lime softening.

2. *Lime-soda softening.* For all forms of calcium hardness. By adding soda ash (Na_2CO_3) non-carbonate hardness is converted to $CaCO_3$ which will then precipitate

$$CaSO_4 + Na_2CO_3 \rightarrow \underline{CaCO_3} + Na_2SO_4$$

As much soda ash is added as there is non-carbonate hardness associated with calcium (Fig. 17.3).

3. *Excess-lime softening.* For magnesium carbonate hardness. The above methods are not effective for the removal of magnesium because magnesium carbonate is soluble

$$Mg(HCO_3)_2 + Ca(OH)_2 \rightarrow \underline{CaCO_3} + MgCO_3 + 2H_2O$$

However, at about pH 11

$$MgCO_3 + Ca(OH)_2 \rightarrow \underline{Mg(OH)_2} + \underline{CaCO_3}$$

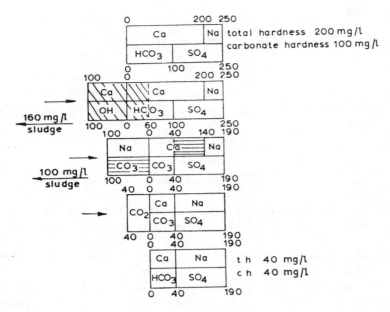

FIG. 17.3. Lime-soda softening.

The practical solubility of $Mg(OH)_2$ is about 10 mg/l. For excess-lime softening it is necessary to add

$$Ca(OH)_2 \equiv HCO_3^-$$
$$+ Ca(OH)_2 \equiv Mg^{++}$$
$$+ 50 \text{ mg/l excess } Ca(OH)_2 \text{ to raise pH.}$$

The high pH level produces good disinfection as a byproduct and thus chlorination may be unnecessary after such softening. Carbonation is necessary to remove the excess lime and reduce the pH after treatment.

4. *Excess-lime soda softening.* For all forms of magnesium hardness. This involves the use of lime and soda ash and is a complicated process.

All forms of precipitation softening produce considerable volumes of sludge. Lime recovery is possible by calcining $CaCO_3$ sludge and slaking with water,

$$CaCO_3 \overset{heat}{\rightarrow} CaO + CO_2$$

$$CaO + H_2O \rightarrow Ca(OH)_2$$

in this way more lime than is required in the plant is produced and the surplus may be sold, the sludge-disposal problem having been solved at the same time.

17.2. Ion Exchange

Certain natural materials, notably zeolites which are complex sodium alumino-silicates and greensands, have the property of exchanging one ion in their structure for another ion in solution. Synthetic ion-exchange materials have been developed to provide higher exchange capacities than the natural compounds.

Ion-exchange treatment has the advantage that no sludge is produced, but it must be remembered that when the ion-exchange capacity has been exhausted the material must be regenerated, which gives rise to a concentrated waste stream of the original contaminant. Industrial wastewaters, such as metal-finishing effluents, can be treated by ion exchange as an alternative to precipitation methods, but again the commonest use of ion exchange is for water softening or demineralization in the case of high-pressure boiler feed waters where high-purity water is essential.

When used for water softening, natural zeolites will exchange their sodium ions for the calcium and magnesium ions in the water, thus giving complete removal of hardness, i.e. representing a zeolite by Na_2X

$$\left.\begin{array}{c} Ca^{++} \\ Mg^{++} \end{array}\right\} + Na_2X \rightarrow \left.\begin{array}{c} Ca \\ Mg \end{array}\right\} X + 2Na^+$$

The finished water is thus high in sodium, which is not likely to be troublesome unless the water was originally very hard. When all sodium ions in the structure have been exchanged, no further removal of hardness occurs. Regeneration can be achieved using a salt solution to provide a high concentration of sodium ions to reverse the exchange reaction,

$$\left.\begin{array}{l} Ca \\ Mg \end{array}\right\} X + 2NaCl \rightarrow Na_2X + \left.\begin{array}{l} Ca \\ Mg \end{array}\right\} Cl_2$$

the hardness being released as a concentrated chloride stream. A natural sodium cycle zeolite will have an exchange capacity of about 200 gram equivalents/m^3 with a regenerant requirement of about 5 equivalents/equivalent exchanged. Synthetic sodium cycle resins may have double the exchange capacity with about half the regenerant requirement, but have a higher capital cost.

Hydrogen cycle cation exchangers produced from natural or synthetic carbonaceous compounds are also available and will also give a water of zero hardness. They exchange all cations for hydrogen ions so that the product stream is acidic and their main use is as the first stage in demineralization operations.

$$\left.\begin{array}{l} Ca^{++} \\ Mg^{++} \\ 2Na^+ \end{array}\right\} + H_2Z \rightarrow \left.\begin{array}{l} Ca \\ Mg \\ 2Na \end{array}\right\} Z + 2H^+$$

Regeneration is by acid treatment:

$$\left.\begin{array}{l} Ca \\ Mg \\ 2Na \end{array}\right\} Z + 2H^+ \rightarrow H_2Z + \left\{\begin{array}{l} Ca^{++} \\ Mg^{++} \\ 2Na^+ \end{array}\right.$$

Typical performance characteristics for a hydrogen cycle exchanger are 1000 gram equivalents/m^3 exchange capacity and a regenerant requirement of about 3 equivalents/equivalent exchanged. Anion exchangers which are usually synthetic ammonia derivatives will accept the product water from a hydrogen cycle exchanger and produce demineralized water for laboratory and other specialized uses as well as for boiler feed water.

A strong anion exchanger ROH, where R represents the organic structure will remove all anions:

$$\left.\begin{array}{l} HNO_3 \\ H_2SO_4 \\ HCl \\ H_2SiO_3 \\ H_2CO_3 \end{array}\right\} + ROH \rightarrow R \left\{\begin{array}{l} NO_3 \\ SO_4 \\ Cl \\ SiO_3 \\ CO_3 \end{array}\right. + H_2O$$

A strong base is necessary for regeneration

$$R \begin{cases} NO_3 \\ SO_4 \\ Cl \\ SiO_3 \\ CO_3 \end{cases} + NaOH \rightarrow R\,OH + Na \begin{cases} NO_3 \\ SO_4 \\ Cl \\ SiO_3 \\ CO_3 \end{cases}$$

Weak anion-exchangers remove strong anions but not carbonates and silicates.

Typical anion-exchanger performance would be 800 gram equivalents/m^3 exchange capacity and 6 equivalents/equivalent exchanged regenerant capacity.

Ion-exchange materials are normally used in units similar to pressure filters and it is possible to combine cation and anion resins in a single mixed-bed unit. The feedwater to an ion-exchange plant should be free of suspended matter, since this would tend to coat the surfaces of the exchange medium and reduce its efficiency. Organic matter in the feed can also cause fouling of the exchanger, although the development of macroporous materials with their internal surface area inaccessible to large organic molecules reduces the problems of organic fouling.

17.3. Adsorption

The presence of trace concentrations of synthetic organic compounds, usually arising from industrial or agricultural operations, in raw water supplies can lead to taste and odour problems in the finished water. Although conventional biological treatment of sewage gives a high removal of organic matter, there is a significant residual of non-biodegradable organic compounds in the final effluent. This residual may give cause for concern in situations where water is abstracted from a receiving water containing considerable amounts of sewage effluent. There is, therefore, increasing interest in the use of adsorption to reduce the concentration of soluble organics in water.

Adsorption is the accumulation of dissolved particles from a solvent on to the surface of an adsorbent. Because adsorption is a surface phenomenon, good adsorbents must have a highly porous structure to provide a large surface area to volume ratio. The most satisfactory adsorbent is activated carbon which is produced from wood ash, lignin, nutshells and similar materials. Dehydration and carbonization are achieved by slow heating in the absence of air and the formation of a highly porous structure (activation) is accomplished by the application of steam, air or carbon dioxide at high temperatures.

When an adsorbent is left in contact with a solution the amount of adsorbed solute increases on the surface of the adsorbent and decreases in the solvent. After some time an adsorption equilibrium is reached when the number of molecules leaving the surface of the adsorbent is equal to the number of molecules being adsorbed on the surface. The nature of the adsorption reaction can be described by relating the adsorption capacity (mass of solute adsorbed per unit mass of absorbent) to the equilibrium concentration of solute remaining in solution. Such a relation is known as an adsorption isotherm (Fig. 17.4).

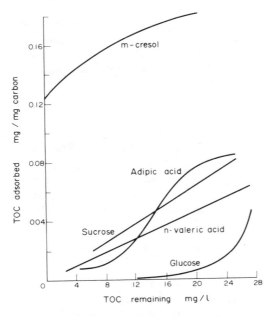

FIG. 17.4. Adsorption isotherms for several organics.

Two simple mathematical models of adsorption are available. The Langmuir isotherm was developed from a theoretical consideration of adsorption based on the concept of equilibrium in a monomolecular surface layer.

$$\frac{x}{m} = \frac{abc}{1 + ac} \tag{17.1}$$

where x = mass of solute adsorbed,

m = mass of absorbent,

c = solute concentration remaining at equilibrium,

a and b are constants.

The Freundlich equation is an empirical relationship which often gives a more satisfactory model for experimental data

$$\frac{x}{m} = kc^{1/n} \tag{17.2}$$

where k and n are constants.

In general, the overall rate of adsorption is governed by the rate of diffusion of the solute into the capillary pores of the adsorbent particle. The rate decreases with increasing particle size and increases with increasing solute concentration and increasing temperature. High molecular weight solutes are not adsorbed as readily as low molecular weight substances and increasing solubility decreases the adsorbability of organic compounds.

When the adsorptive capacity has been fully utilized, the carbon must be discarded if in the form of powder or regenerated by heat treatment if in the more costly granular form. The use of powdered carbon is usually restricted to situations which require intermittent or occasional removal of low concentrations of organic matter. Where continuous removal of significant concentrations of organic substances is necessary, granular carbon in beds, possibly with on-site regeneration, would be required.

Further Reading

BURKE, T., HYDE, R. A. and ZABEL, T. F., The performance and cost of activated carbon for control of organics. *J. Instn Wat. Engrs Scits*, **35**, 1981, 329.

ECKENFELDER, W. W., Jr., *Industrial Water Pollution Control*, McGraw-Hill, New York, 1966, Chapters 6, 7 and 8.

HOLDEN, W. S. (Ed.), *Water Treatment and Examination*, Churchill, London, 1970, Chapters 34 and 35.

MARTIN, R. J. and IWUGO, K. O., Studies on residual organics in biological plant effluents and their treatment by the activated carbon adsorption process. *Pub. Hlth Engr*, 7, 1979, 61.

Problems

1. Make up a bar diagram in terms of calcium carbonate for a water with the following composition:

Ca^{++}	101 mg/l
Mg^{++}	4.75 mg/l
Na^+	14.0 mg/l
HCO_3^-	220 mg/l
$SO_4^=$	88.4 mg/l
Cl^-	21.3 mg/l

   Ca 40, Mg 24.3, Na 23, H 1, C 12, O 16, S 32, Cl 35.5. (Total hardness 271.6 mg/l, carbonate hardness 180 mg/l)

2. Soften the water in question 1 by lime and lime-soda treatment and determine the final hardness of the water in each case. (Total hardness 131.6 mg/l and 59.6 mg/l. Carbonate hardness 40 mg/l in both cases)

3. A sodium cycle cation exchanger has a volume of 10 m³ and an exchange capacity of 400 gram equivalents/m³. Determine the volume of water of initial hardness 250 mg/l as $CaCO_3$ which can be softened in the exchanger. If regeneration requires 5 equivalents/equivalent exchanged determine the amount of sodium chloride necessary for regeneration. (800 m³, 1.17 tonne)

4. The following data were obtained from laboratory tests using powdered activated carbon:

Carbon dose mg/l	Initial TOC mg/l	Final TOC mg/l
12	20	5
7	12	2

   Determine the carbon dose necessary to reduce an initial TOC of 15 mg/l to 3 mg/l assuming that the Langmuir isotherm applies. (9.1 mg/l)

CHAPTER 18

Sludge Dewatering and Disposal

ONE of the major problems in water and wastewater treatment is that of sludge disposal. Large volumes of sludge with high water contents are produced from sedimentation tanks and their dewatering and ultimate disposal may account for as much as half the cost of treatment.

18.1. Principles of Dewatering

The types of sludge which are produced in treatment processes are:
1. Primary sludge from wastewater sedimentation.
2. Secondary sludge from biological wastewater treatment.
3. Digested forms of the above separately or mixed.
4. Hydroxide sludges from coagulation and sedimentation of waters and industrial wastes.
5. Precipitation sludges from softening plants and from industrial waste treatment.

All of these sludges have low solids contents (1–6%) and thus large volumes of sludge must be handled to dispose of even a relatively small mass of solids. The main concern in treatment of sludge is therefore to concentrate the solids by removing as much water as possible. The density and nature of the solid particles have a considerable influence on the thickness of the sludge produced. Thus a metallurgical ore slurry with solids of S.G. 2.5 will quickly separate to give a sludge with a solids content of about 50%. On the other hand, sewage sludge contains highly compressible solids with S.G. about 1.4 and will only produce a sludge of 2–6% solids. Unless care is exercised, attempts to increase the solids content by draining off excess water may cause the solids to compress, thus blocking the voids and preventing further drainage. Primary sewage sludge has a heterogeneous nature with fibrous solids so that drainage is easier than from digested sludge which is much more homogeneous in nature.

The S.G. of a sludge with a particular solids content can be determined from the following expression:

$$\text{sludge S.G.} = \frac{100}{\left(\dfrac{\%\ \text{solids}}{\text{solids S.G.}}\right) + \left(\dfrac{\%\ \text{water}}{\text{water S.G.}}\right)} \qquad (18.1)$$

The influence of a reduction in moisture content on the volume occupied by a sludge can be shown by a simple example. Consider a sludge with 2% solids of S.G. 1.4 : 20 kg of solids are accompanied by 980 kg of water, thus the sludge S.G. is 1.006 and the volume occupied is 0.994 m^3. By dewatering to a solids content of 25% (a soil-like consistency) the 20 kg of solids are now accompanied by only 60 kg of water so that the sludge S.G. becomes 1.077 and the volume occupied reduces to 0.074 m^3, i.e. about 7.5% of the original volume.

The work of Carman on filtration as stated in equation (13.3) was developed by Coackley[1] for the dewatering of sludges by filtration. The parameter specific resistance was introduced to compare the filtrability of different sludges.

The rate of filtration, i.e. the ease of dewatering, is given by

$$\frac{dV}{dt} = \frac{PA^2}{\mu(rcV + RA)} \qquad (18.2)$$

where V = filtrate volume obtained after time t,
P = applied pressure,
A = filter area,
μ = absolute viscosity of filtrate,
r = specific resistance of sludge,
c = solids concentration of sludge,
R = resistance of clean filter medium.

For constant P integration gives

$$t = \frac{\mu rc}{2PA^2}V^2 + \frac{\mu R}{PA}V \qquad (18.3)$$

or

$$\frac{t}{V} = \frac{\mu rc}{2PA^2}V + \frac{\mu R}{PA} \qquad (18.4)$$

Using a laboratory filtration apparatus it is possible to determine the specific resistance by plotting t/V against V, the slope of the line being $\mu r c/2PA^2$. The higher the value of the specific resistance the more difficult the sludge will be to dewater.

An alternative method of assessing the filtrability of a sludge is to measure the capillary suction time (CST) devised by Baskerville and Gale.[2] The CST depends on the suction applied to a sample of sludge by an absorbent chromatography paper. An area of paper in the centre is exposed to the sludge whilst the remaining paper is used to absorb the water removed from the sludge by capillary suction. The time taken for water to travel a standard distance through the paper is noted visually or electronically and is found to have good correlation with specific resistance values for a particular sludge.

18.2. Sludge Conditioning

To improve the efficiency of the dewatering process it is often useful to provide a preliminary conditioning stage to release as much bound water as possible from the sludge particles, to encourage solids agglomeration and increase the solids content. Various methods of conditioning are employed depending upon the characteristics of the sludge to be treated.

1. *Thickening.* With many flocculent sludges, particularly surplus activated sludge, slow-speed stirring in a tank with a picket-fence type mechanism encourages further flocculation and can significantly increase the solids content and settleability, allowing supernatant to be drawn off.

2. *Chemical conditioning.* Chemical coagulants can be useful in promoting agglomeration of floc particles and releasing water. Common coagulants used for sludge conditioning are: aluminium sulphate, aluminium chlorohydrate, iron salts, lime and/or polyelectrolytes. The cost of these reagents is usually more than covered by the increased solids content and improvement in dewatering characteristics arising from their use.

3. *Elutriation.* The chemical requirement for conditioning can be reduced by mixing the sludge with water or effluent and allowing settlement and removal of supernatant to take place before chemicals

are added. This washing process removes much of the alkalinity which in digested sludges exerts a high chemical demand.

4. *Heat treatment.* A number of processes have been employed to heat wastewater sludges under pressure with the aim of stabilizing the organic matter and improving dewaterability. A typical operation involves heating to a temperature of about 190 °C for 30 min at a pressure of 1.5 MPa, the sludge being then transferred to thickening tanks. The supernatant has a high soluble organic content and must be returned to the main oxidation plant for stabilization which is not always easy due to its limited biodegradability. Corrosion problems and high energy costs have meant that heat treatment plants are not now very attractive in most situations.

18.3. Sludge Dewatering

For many sludge-disposal methods preliminary dewatering is essential if the costs of disposal are to be kept under control and a variety of dewatering methods are employed (Fig. 18.1) depending upon land availability and the costs related to the particular situation.

Drying Beds

The oldest and simplest dewatering process uses shallow rectangular beds with porous bottoms above an underdrain layout. The beds are divided into convenient areas by low walls. Sludge is run on to the beds to give a depth of 125–250 mm and dewatering takes place due to drainage from the lower layers and evaporation from the surface under the action of sun and wind. The cake cracks as it dries allowing further evaporation and the escape of rainwater from the surface. In good conditions a solids content of around 25 % can be achieved in a few weeks but a more normal period in temperate climates would be 2 months. Best results are obtained by applying shallow layers of sludge frequently rather than deep layers at longer intervals. Removal of dried sludge can be undertaken manually at small works but elsewhere mechanical sludge-lifting plant must be used. A typical area requirement for sewage sludge is 0.25 m²/head of population and it is this large requirement which makes the feasibility of drying beds dubious unless

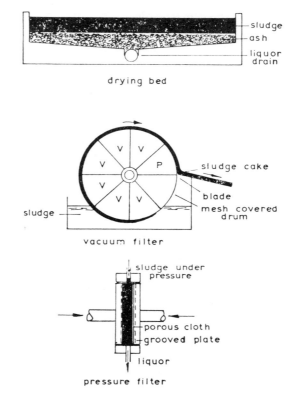

Fig. 18.1. Dewatering methods.

land is availabe at low cost. In many circumstances some form of mechanical dewatering, for which land requirements are small and in which the performance is not affected by the weather, is utilized.

Pressure Filtration

Pressure filtration is a batch process in which conditioned sludge is pumped with increasing pressure into chambers lined with filter cloths which retain the solids but allow liquid to escape via grooves in the metal backing plates. As liquid escapes, the cake formed adjacent to the cloth acts as a further filter for the remainder of the sludge so that the cake dewaters towards the centre. Pressing times vary from 2–18 h with pressures of 600–850 kPa giving a resultant cake solids content of 25–50%. Solids loading depends upon the nature of the sludge and length of pressing cycle.

A development of pressure filtration in the form of the belt press provides continuous operation by introducing conditioned sludge into the gap between two endless belts to which pressure is applied by means of rollers. Dewatering occurs by a combination of gravity drainage, pressure filtration and shear.

Vacuum Filtration

This is a continuous process in which a revolving segmented drum covered with filter cloth is partially submerged in conditioned sludge. A vacuum of about 90 kPa is applied to the submerged segments and sludge is attracted to the surface of the cloth. As the drum rotates and the layer of sludge emerges from the tank, air is drawn through it by the vacuum to assist dewatering. A scraper blade removes the sludge cake assisted by change to positive pressure in the relevant drum segment. The cake solids are usually 20–25% with filter yields of around 20 kg dry solids/m² h.

Centrifugation

Continuously operated centrifuges have some application for sludge dewatering. Most are of the solid-bowl type in which conditioned sludge is fed into the centre of a rapidly rotating bowl. The solids are thrown to the outer edge of the bowl from where they are removed by a scraper/conveyor. Centrifuges are relatively compact but are not usually able to achieve solids concentrations greater than 20% and in many cases it is difficult to economically reach solids contents higher than about 12–15% from water and wastewater sludges.

It is important to appreciate that in all sludge dewatering operations the separated liquid requires suitable disposal arrangements. With wastewater sludges the liquid is usually highly polluting and must be returned to the main treatment plant for stabilization.

18.4. Sludge Disposal

Sewage sludge has some value as a soil conditioner since it contains significant amounts of nitrogen and phosphorus and thus sludges derived from works without major industrial discharges may be readily disposed of to farm land. Problems can arise from sludges containing such potentially toxic materials as heavy metals so that use of agricultural land for disposal may not be possible. Prior anaerobic digestion of sludges before land disposal is often required to ensure destruction of any pathogenic micro-organisms which might be present in the raw sludge. Application of sludge to agricultural land is often made in the wet state by tanker vehicle or spray irrigation system but climatic conditions may restrict the application at certain times so that sufficient storage must be provided to cover these situations. Depending upon agricultural practices, soil type and climate, a land disposal area of $20\,m^2$/person upward is necessary.

A good deal of sludge is dumped in old quarries and similar structures and by land filling with dewatered sludge it is possible to reclaim waste areas. The possibility of groundwater contamination as the result of such practices must always be borne in mind. In coastal areas the disposal of liquid sludge at sea in suitable deep-water sites is fairly common and there is little evidence of any significant environmental damage caused by such actions if properly controlled. However, some regulatory authorities now ban the disposal of wastewater sludges at sea.

The composting of sewage sludge with domestic refuse produces a stable material with good soil-conditioning properties but with little reduction in volume so that composting is only justifiable if there is a market for the product. Mixing all the sewage sludge and all the refuse from a community produces a mixture too wet for composting to occur and it is necessary to dewater the sludge to about 25% solids for the process to operate satisfactorily.

In situations with no sludge-disposal sites within economic reach or where the sludge contains toxic matter the best solution may be that of incineration. This is usually carried out in multiple-hearth furnaces and the incineration is normally self sustaining when sewage sludges are dewatered to around 25% solids (sewage solids have a calorific value which is typically 20 MJ/kg). After incineration the residual ash amounts to only 5–10% of the original solids thus greatly reducing the disposal problem but involving large capital expenditure and considerable operating expense.

The volume and nature of sludges produced by water-treatment operations are such that their disposal does not normally cause great problems. Land application, lagooning or filter pressing with disposal to a tip are common practices. Aluminium hydroxide sludges have no fertilizer value but lime sludges from softening operations have considerable agricultural value.

References

1. COACKLEY, P., Research on sewage sludges carried out in the Civil Engineering Department of University College, London. *J. Proc. Inst. Sew. Purif.* 1955 (1), 59.
2. BASKERVILLE, R. C. and GALE, R. S., A simple automatic instrument for determining the filtrability of sewage sludges. *Wat. Pollut. Control*, **67**, 1968, 233.

Further Reading

BEST, R., Want not, waste not! Sensible sludge recycling. *Wat. Pollut. Control*, **79**, 1980, 307.
HAMLIN, M. J. and EL HATTAB, I., Factors affecting the dewatering of sludge. *J. Instn Pub. Hlth Engrs*, **66**, 1967, 101.
HUDSON, J. A. and FENNEL, H., Disposal of sewage sludge to land: Chemical and microbiological aspects of sludge to land policy. *Wat. Pollut. Control*, **79**, 1980, 370.
INSTITUTION OF WATER ENGINEERS, Disposal of waterworks sludge: Final report of Research Panel No. 14. *J. Instn Wat. Engrs*, **27**, 1973, 399.
METCALF and EDDY INC., *Wastewater Engineering: Treatment Disposal Reuse*, McGraw-Hill, New York, 1979, Chapter 11.
VESILIND, P. A., *Sludge and its Disposal*, Ann Arbor Science Publishers, Ann Arbor, 1979.
WILLIAMS, J. H., Use of sewage sludge on agricultural land and the effects of metals on crops. *Wat. Pollut. Control*, **74**, 1975, 635.
WILLIAMSON, D. J. and WHEALE, G., The influence of sludge storage on chemical conditioning costs. *Wat. Pollut. Control*, **80**, 1981, 529.

Problems

1. A sewage flow of 1 m^3/s containing 450 mg/l SS is given primary sedimentation to remove 50% of the SS before discharge to the sea. Calculate the daily volume of sludge produced if it is drawn off from the sedimentation tanks at 4% solids. Assume solids have S.G. 1.4. (480 m^3)

2. Given a volume of $100 \, m^3$ of sludge at 95% moisture content determine the volume occupied by the sludge when dewatered to 70% moisture. Assume solids have S.G. 1.4. $(15.5 \, m^3)$

3. The following results were obtained from a specific resistance determination on a sample of activated sludge:

Time, s	Volume of filtrate, ml
0	0.0
60	1.4
120	2.4
240	4.2
480	6.9
900	10.4

Vacuum pressure	97.5 kPa
Filtrate viscosity	$1.011 \times 10^{-3} \, Ns/m^2$
Solids content	$7.5\% \, (75 \, kg/m^3)$
Area of filter	$4.42 \times 10^{-3} \, m^2$

Plot the values of t/V against V and hence obtain the slope and calculate the specific resistance. $(2.4 \times 10^{14} \, m/kg)$

CHAPTER 19

Tertiary Treatment and Water Reclamation

EVEN in their 1912 report the Royal Commission on Sewage Disposal envisaged the need for effluent standards more stringent than the norm of 30 mg/l SS and 20 mg/l BOD. In situations where the receiving water affords little dilution to the effluent discharge and/or where there is raw water-abstraction downstream, it has become common practice to require an effluent of higher standard than 30:20. Similar requirements may apply when effluents are used for industrial water-supply purposes in lieu of higher grade raw waters. When this higher quality discharge is essentially achieved by removal of suspended matter and its associated BOD it has become convention to use the term tertiary treatment.

Because of increasing concern at the presence of soluble non-biodegradable organics, inorganic nutrients and salts and trace contaminants of various types in conventionally treated effluents there is now considerable interest in advanced waste treatement or water-reclamation techniques. Such techniques may have applications in eutrophication control, water re-use for potable supply and of course in the production of freshwater from saline sources.

19.1. Tertiary Treatment

Although a conventional sewage-treatment plant incorporating primary settlement, biological oxidation and final settlement may be able to produce an effluent of better than 30:20 standard at times, the reliable production of an effluent significantly better than 30:20 requires some form of tertiary treatment. The Royal Commission recommendation that 4 mg/l BOD be the limiting level of pollution in receiving waters to prevent undesirable conditions has been found to be rather unrealistic since there are many rivers in the UK which have BOD levels in excess of 4 mg/l but which have high DO levels and are used for water-supply purposes. The need for tertiary treatment in a particular situation should therefore be assessed in the light of the circumstances relevant to that situation, i.e. dilution, reaeration characteristics, downstream water use, etc.

The main reason for limiting SS in effluents is that they may settle on the stream bed and inhibit certain forms of aquatic life. Flood flows may resuspend these bottom deposits and exert sudden oxygen demands. Settlement does not, however, always occur in waters which are naturally turbid. Effluent SS levels may not be of great significance in themselves, although they do of course influence the BOD of the effluent. In some cases where a restriction on SS levels is desirable a reduction in BOD to below 20 mg/l may also be necessary. This is not always so, however, and the need for BOD reduction should be examined separately. The BOD is by its very nature a somewhat unreliable parameter and BOD standards should not be specified with excessive accuracy. Possible BOD standards could be 20, 15, 10 and exceptionally 5 mg/l.

Most forms of tertiary treatment used in the UK have been aimed at removal of some of the excess SS in the effluent from a well-operated conventional works. Tertiary treatment should be considered as a technique for improving the quality of a good effluent and not as a method of trying to convert a poor effluent into a very good-quality discharge. Removal of SS from an effluent gives an associated removal of BOD due to the BOD exerted by the suspended matter. There is a good deal of evidence to show that for normal sewage effluents the removal of 10 mg/l SS is likely to remove about 3 mg/l BOD.

Rapid Filtration

Various methods of tertiary treatment are available, but the method which finds most application in large works is that of rapid gravity filtration. Most installations are based on the downflow sand filter which has been used in water-treatment plants for many years. More efficient forms of filter, including mixed media beds and upflow units, have been used with some success, but in many cases the downflow unit is adopted because of its simplicity. The variable nature of the SS present in the effluent from final settling tanks makes prediction of the performance of any tertiary treatment unit difficult. Because of the wide variation in filtration characteristics of suspended matter it is always advisable to carry out experimental work on a particular effluent before proceeding with design work.

It is generally assumed that rapid gravity filters operated at a hydraulic loading of about $200 \, m^3/m^2 \, d$ should remove $65-85\%$ SS and $20-35\%$ BOD from a $30:20$ standard effluent. Because of the relatively short time which elapses between backwashes ($24-48 \, h$ generally), little biological activity occurs and thus rapid filters are not likely to achieve any significant oxidation of ammonia. The SS removal is not significantly affected by the hydraulic loading within quite wide variations and there is little benefit in using sand smaller than $1.0-2.0 \, mm$ grading.

Slow Filtration

On small works slow sand filters are sometimes employed for tertiary treatment at loadings of $2-5 \, m^3/m^2 \, d$. Slow filters have low operation and maintenance costs, but their relatively large area requirements normally rule them out for other than small installations. They can be expected to remove $60-80\%$ SS and $30-50\%$ BOD. Slow filters provide a significant amount of biological activity, thus encouraging BOD removal and providing a degree of nitrification.

Microstraining

Microstrainers have been utilized for tertiary treatment since 1948 and a considerable number of installations are in operation. They have the advantage of small size and can thus be easily placed under cover. Removals of SS and BOD depend upon the mesh size of the fabric used and the filterability characteristics of the suspended matter. Reported removals range from $35-75\%$ SS and $12-50\%$ BOD. Microstraining should reliably give an effluent of $15 \, mg/l$ SS and $10 \, mg/l$ SS should be possible with a good final tank effluent. Biological growths on the fabric which could cause clogging and excessive head loss can normally be controlled by UV lamps.

Upward-flow Clarifier

This technique was originally developed as a means of obtaining better quality effluents from conventional humus tanks on small bacteria bed installations. The process is based on passing the tank effluent through a 150 mm layer of 5–10 mm gravel supported on a perforated plate near the top of a horizontal flow humus tank with surface overflow rates of 15–25 m³/m² d. Passage through the gravel bed causes flocculation of the suspended matter and the floc settles on top of the gravel. Accumulated solids are removed intermittently by drawing down the liquid level below the gravel bed. Removals of 30–50 % SS can be achieved dependent upon the size of gravel and the type of solids. Similar results have been achieved with wedge wire in place of gravel and there is evidence that many types of porous materials can be used to promote flocculation.

Grass Plots

Land irrigation on grass plots can provide a very effective form of tertiary treatment particularly suitable for small communities. Effluent is distributed over grassland, ideally with a slope of about 1 in 60, and collected in channels at the bottom of the plot. Hydraulic loadings should be in the range 0.05–0.3 m³/m² d and 60–90 % SS and up to 70 % BOD removals can be achieved. Short grasses are preferable, but the sowing of special grass mixtures does not appear to be worthwhile. The area should be divided into a number of plots to permit access, for grass and weed cutting—growth is likely to be prolific due to the nutrients present in effluents.

Lagoons

Storage of effluent in lagoons provides a combination of sedimentation and biological oxidation depending upon the retention time. With short retention times (2–3 d) the purification effect is mainly due to flocculation and sedimentation, with SS removals of 30–40 % being likely.

With longer retention times (14–21 d) the improvement in quality may be very marked with 75–90 % SS, 50–60 % BOD and 99 % coliform removals. Heavy algal growths may, however, occur with these larger ponds, and at times the escape of algae from the pond can result in relatively high SS and BOD levels in the final discharge. The improvement in bacteriological quality in lagoons is greater than provided by most other forms of tertiary treatment, with the possible exception of grass plots, and is of particular interest if the receiving water is a raw water source. In UK conditions it would appear that a retention time of about 8 d provides the most satisfactory overall performance, since longer retention times are more likely to give rise to excessive algal growths.

19.2. Water Reclamation

Increasing demands for water will in the future require the development of new sources, some of which will contain water of a quality inferior to that judged acceptable in the past for water-supply purposes. In densely populated areas, much of the increased demand may have to be satisfied by abstractions from lowland rivers which are likely to contain significant proportions of sewage effluent and industrial wastewaters.

The direct re-use of sewage effluent to satisfy a number of industrial water requirements is already an accepted practice, with consequent economies in cost and one which also serves to release supplies of better grade water which would otherwise be used industrially. The use of sewage as a source of potable water is technically feasible at the present time, but would be relatively costly and its use would probably produce psychological objections from the consumers. Such direct re-use would require the adoption of additional processes mainly physico-chemical in nature, which are likely to be fairly costly. The adoption of such techniques should therefore only be as the result of careful cost−benefit analysis of the situation.

Conventional water treatment (coagulation, sedimentation, filtration and disinfection) was originally developed for the removal of suspended and colloidal solids from raw waters together with limited removal of the soluble organics responsible for the natural colour in water from upland catchments. Certain soluble constituents, such as those responsible for hardness, can be removed by the incorporation of precipitation or ion-exchange processes. Using such techniques it is possible to produce an acceptable water from relatively heavily polluted sources, but it must be appreciated that there are limits to the levels of certain types of impurity which can be satisfactorily dealt with by conventional water treatment. Indeed some impurities may be completely unaltered by normal water-treatment methods. Nevertheless, there are a number of examples of situations where emergency conditions have required the treatment of heavily polluted waters without danger to the consumer. In such circumstances it may be necessary to distinguish between potable and wholesome supplies, since it is by no means certain that a potable supply would also be considered wholesome.

Temporary use of the River Avon at Ryton[1] is an illustration of the results (Table 19.1) of subjecting a raw water from a river heavily polluted by sewage to conventional water treatment. The results illustrate the limitation of conventional treatment as regards the removal of dissolved solids, chloride, nitrogen compounds and dissolved organics. Later stages of the abstraction in Ryton involved the addition of a high-rate bacteria bed ahead of the main plant. This nitrified ammonia which reduced disinfection problems and was also useful in oxidizing organic matter responsible for taste and odour troubles.

TABLE 19.1. RIVER AVON SUPPLY[1]

Characteristic (mg/l except where noted)	Raw water		Treated water	
	Average	Range	Average	Range
Turbidity (units)	25	5–425	0.6	0–7.6
Colour (°H)	21	12–70	4	0–16
TDS	768	516–1012	759	532–968
Cl	48	29–68	50	33–69
Amm.N	1.4	0.6–17.4	1.3	0–17.4
Alb.N	0.55	0.35–1.3	0.47	0.11–0.92
NO_3–N	7	2–10	8	3–13
PV	3.7	2.7–5.1	1.9	0.8–2.7
E. coli/100 ml	9700	200–25 000	0	0–0
37°C colonies/ml	5100	20–29 000	4	0–20
22°C colonies/ml	87 000	2400–600 000	11	1–60

A more direct form of effluent re-use took place for a short period in 1956–7 at Chanute in the USA[2] where drought conditions caused a failure of the water supply and sewage effluent was recycled into the water-treatment plant. Both treatment plants were of conventional design and at the end of a 5-month period, when it was estimated that about ten cycles had taken place, it was felt that the limit had been reached. Table 19.2 gives an indication of the build-up of impurities during the operation and again illustrates the inability of conventional treatment processes to deal with dissolved solids, nitrogen compounds and most dissolved organics.

TABLE 19.2. REUSE AT CHANUTE[2]

Characteristic mg/l	Original water	After ten cycles
TDS	305	1139
Cl	63	520
Amm.N	—	10
NO_3–N	1.9	2.7
COD	—	43
ABS	—	4.4
SO_4	101	89
PO_4	—	3.9

Thus although it is possible to obtain a marked improvement in the quality of sewage effluent by the application of conventional water-treatment processes, it seems unlikely that the product water would be acceptable as a potable supply on a regular basis. It should be noted, however, that simple coagulation of sewage effluent can produce a water which would be suitable for many industrial purposes.

It will be clear from the foregoing comments that to obtain a wholesome potable supply of water from heavily polluted sources such as sewage effluent, conventional water treatment alone is not sufficient. To achieve the desired end quality a number of alternative courses of action could be adopted.

(i) Provide additional treatment stages at sewage and/or water-treatment plants to deal with contaminants not affected by normal treatment.

(ii) Provide a completely new form of sewage and/or water treatment.

(iii) Use conventional treatment processes and blend the finished water with another water of higher quality so that the mixture is of acceptable quality.

(iv) Dispense with separate sewage and water-treatment facilities (and the intervening receiving water) and introduce an integrated water reclamation facility.

Courses (iii) and (iv) are certainly feasible from a technological point of view at the present time. The use of distillation or reverse osmosis would permit the production of an acceptable water at a cost similar to that of producing fresh water from sea water. Such techniques could permit the direct recycling of sewage effluent, although a certain amount of make-up water would be necessary because of losses in the system. Completely closed-cycle systems for water and sewage are of considerable interest in space-vehicle development, but in this application costs are likely to be of secondary importance. It would thus seem likely that for large-scale use of sewage as a raw water source alternatives (i) or (iii) above are most likely to have the greatest application.

The continuance of the system by which sewage is treated and discharged to a receiving water from which abstractions may be made for water-supply purposes has much to commend it. The presence of dilution water is useful (assuming that the receiving water is of better quality than the effluent) and the concentration of non-conservative pollutants will be reduced by self-purification between discharge and abstraction points. Consumer acceptance of such an indirect recycling scheme is also likely to be better than for direct recycling. In this context it is necessary to consider the characteristics of conventionally treated sewage effluent which would be undesirable in water-treatment terms. Table 19.3 shows typical analyses of crude sewage and sewage effluents compared with EEC recommendations. Examination of these analyses indicates that a number of the characteristics of sewage effluent are likely to be unacceptable as regards treatment in a conventional water works. In particular, the high contents of non-biodegradable organics, total solids, ammonia and nitrate nitrogen are likely to pose some problems in treatment. In addition, the very high levels of bacteriological (and viral) impurity would be such as to cause concern to water-treatment authorities.

Dissolved organics, which are largely non-biodegradable or, at least, only slowly broken down by biological means, are found in sewage effluents in the form of many different compounds whose presence is indicated by high TOC and COD values with relatively low BOD values. Some of the organics in sewage effluents are believed to be similar to the

TABLE 19.3. TYPICAL SEWAGE CHARACTERISTICS AND WATER QUALITY LIMITS

Characteristic	Crude sewage	Conventional effluent (30:20 std)	Sand filtered effluent	EEC raw water directive for A3 treatment (see Table 2.4)	
				Guide limit	Mandatory limit
BOD mg/l	400	20	10	7	—
COD mg/l	800	100	70	30	—
BOD/COD ratio	0.50	0.20	0.14	0.23	—
TOC mg/l	300	35	25	—	—
SS mg/l	500	30	10	—	—
Turbidity (units)	—	—	—	—	—
Cl mg/l	100*	100*	100*	200	—
Org.N mg/l	25	0	0	3	—
Amm.N mg/l	25	5†	4†	2	4
NO₃–N mg/l	0	20†	21†	—	11.3
Total solids mg/l	1000*	1000*	1000*	(1000 μS/cm)	—
Total hardness mg/l	250*	250*	250*	—	—
PO₄ mg/l	10	6	6	0.7	—
ABS mg/l	2	0.2	0.2	0.5	—
Colour (°H)	—	50	30	50	200
Coliform MPN/100 ml	10⁷	10⁶	10⁵	5 × 10⁴	—

* Concentration depends to some extent on quality of carriage water.
† Depends upon degree of nitrification achieved in sewage treatment.

compounds which produce natural colour in upland catchments. It is, however, certain that other organic compounds may be more troublesome in water, particularly as regards the formation of tastes and odours. They are also likely to increase the chlorine demand and the possibility of toxic effects must also be considered.

Total solids is also a somewhat vague parameter for the assessment of water quality, since it gives no indication of the source or nature of the impurities. Clearly, a few mg/l of certain compounds could be toxic, whereas several hundred mg/l of other compounds would be quite harmless. There is unfortunately a great lack of information about the effects on man of inorganic compounds in water. In situations where a water is within normal potable standards with the exception of total solids there is often a willingness to accept a total solids level higher than that considered desirable by WHO.

Nitrogen compounds in the form of ammonia or nitrate are present in all sewage effluents and are undesirable in potable water because of disinfection problems and promotion of biological growths with resulting tastes and odours (ammonia) and due to the potential health hazard to young babies from nitrates.

Phosphates, another normal constituent of sewage effluent, may also be troublesome in water-treatment processes because of the inhibiting effect they can sometimes produce on coagulation reactions.

The existence of large numbers of micro-organisms in a raw water is always of great concern to water-treatment authorities, but since disinfection is a well-established process there seems no reason why bacteriologically satisfactory water could not be produced from sewage effluent. The inactivation of viruses is, however, somewhat less predictable and thus their presence in raw water is particularly undesirable.

19.3. Removal of Impurities not Amenable to Conventional Treatment

Dissolved Organics

Biological oxidation of polluted raw water on a high-rate bacteria bed can achieve some reduction in BOD, possibly of about 20%. Such installations are, however, usually installed primarily for the oxidation of ammonia. More substantial removal of soluble organics can be achieved by adsorption on activated carbon. Both powdered and granular forms of activated carbon can be employed to give relatively high removals of COD and TOC, although a residual of unadsorbable material may remain. For intermittent use, powdered carbon may be satisfactory with the addition being made to the coagulation/sedimentation stage or to the filters. Where continuous use of activated carbon is necessary the granular form is more appropriate and provision must be made for regeneration either on-site or in a central facility. As well as reducing COD and TOC levels, activated carbon treatment is usually also able to give significant reductions in the colour and in the taste and odour of waters. It must be appreciated, however, that activated carbon adsorption does not provide a solution for all situations in which organic contamination creates problems.

Dissolved Solids

Most forms of treatment have little or no effect on the total dissolved solids content of water so that in a re-use situation the accumulation of dissolved solids may well limit the number of cycles which are possible. In many parts of the world, brackish groundwaters are found with TDS levels in excess of those acceptable in potable supplies and in arid areas these groundwaters or seawater may be the only available source of water. There has thus been considerable interest in the development of processes for the removal of excessive dissolved solids from such sources and these processes also have some application in the removal of dissolved solids from sewage and industrial effluents. The scale of the problem is somewhat different in that seawater has a TDS level of around 35 000 mg/l whereas the TDS levels of most effluents are around 1000 mg/l.

The distillation of seawater in evaporators has long been an accepted procedure for obtaining high-purity water although the finished product is not acceptable for drinking water until it has been aerated and chemically treated. The capital and operational costs for distillation are very high so that the process is normally only used in situations where alternative sources of water are unavailable. Modern distillation plants operate on the multi-stage flash process (Fig. 19.1) which is based on the preheating of a pressurized salt-water stream, composed of seawater feed and return brine solution, with final heating taking place in a steam-fed heat exchanger. The heated salt water is then released to the first chamber where pressure is reduced thus allowing part of the water to flash vaporize to steam which is condensed in a heat exchanger fed with the incoming salt solution. The remaining salt water passes to the next stage which operates at a slightly lower pressure so that further evaporation occurs. Most plants have thirty to forty stages with a temperature range of ambient $+5°C$ to $110°C$. Distillation costs depend upon the size of the plant and whether some of the steam can be used for electricity generation to partly offset production costs. The basic energy requirement for multi-stage flash distillation is about 200 MJ/m^3 of distillate.

In contrast to the complexity of flash-distillation installations, the use of solar stills has been studied in some parts of the world. These are low technology devices with a free source of energy but due to their low yield

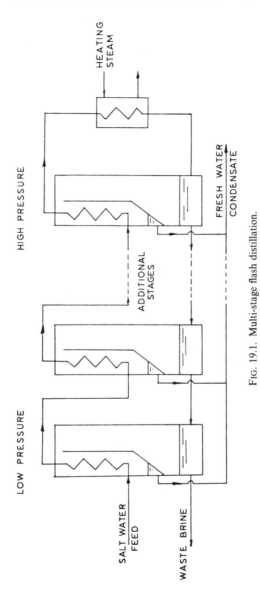

FIG. 19.1. Multi-stage flash distillation.

and the relatively high cost of the necessary glass structures the actual costs of water production are not greatly different from those of large flash distillation plants.

Reverse osmosis (Fig. 19.2) depends upon the phenomenon of osmosis in which certain types of membrane will permit the passage of fresh water whilst preventing or restricting the movement of soluble materials. Thus, if a semi-permeable membrane is used as a barrier between a salt solution and freshwater the solvent (i.e. the water) will pass through the membrane to equalize the salt concentrations on either side. This movement occurs because of the osmotic pressure exerted by the dissolved salt. The process can be thought of as a form of hyper-filtration in which water molecules are small enough to pass through the pores in the membrane but larger molecules are unable to do so. The osmotic pressure is directly proportional to the concentration of the solution and to the absolute temperature and for sea water the osmotic pressure is about 2.5 MPa. If a salt solution is subjected to a pressure greater than its osmotic pressure, water will pass through the membrane giving a desalted product and leaving a concentrated brine. In practice, to obtain a significant yield of desalted seawater it is necessary to operate at pressures of 4–6 MPa and even then yields are only of the order of $0.5–2.5 \, m^3/m^2 \, d$. Membranes are usually made from a cellulose acetate or polyamide base and most commercial units utilize tubular systems to support the membranes at the

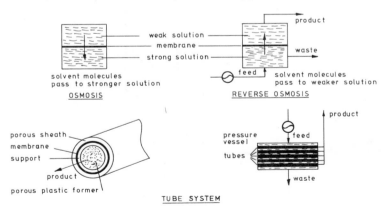

FIG. 19.2. Principles of reverse osmosis.

high pressures necessary for economic operation. Modern membranes can give 99 % salt rejection so that a product of less than 500 mg/l TDS can be obtained from seawater in a single pass. Energy requirements for reverse osmosis are around $60-100 \, MJ/m^3$ of distillate but the high cost of the membranes, which have a limited life, means that the final cost of reverse osmosis-produced water is not far removed from that of flash distillation when used for seawater. In order to protect membrane life it is usually necessary to provide conventional water treatment before the supply is fed to the reverse osmosis units.

For brackish groundwater the electrodialysis process (Fig. 19.3) is fairly widely used although it is not suitable for handling seawater or supplies of similar salinity. Batteries of ion-selective membranes are placed in a cell so that when an electrical potential is applied migration of ions occurs, giving alternate reduced-salinity and concentrated-salinity chambers. Pre-treatment of the raw water is not so important as with the reverse osmosis process although organic matter and sulphates can cause fouling of the membranes and iron and manganese should be removed to prevent their precipitation on the membrane surfaces.

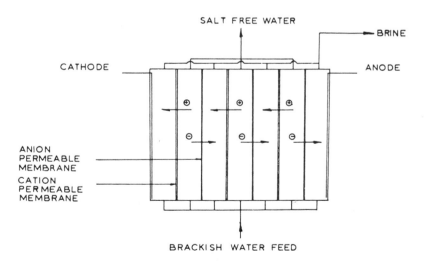

FIG. 19.3. Electrodialysis.

On thermodynamic considerations, desalination by freezing compares favourably with distillation and, on an operational basis, low temperatures should give rise to fewer problems with corrosion and scaling. However, although a great deal of experimental work has been undertaken with freezing plants many practical problems have arisen and the future of the process does not now seem very promising.

All desalination processes produce a waste-brine stream for which suitable disposal arrangements must be made.

Nitrogen Compounds

The amount of ammonia present in sewage effluent can be reduced to low levels by biological nitrification either at the sewage works before discharge to the receiving water, which will prevent fish toxicity problems, or by pretreatment at the water works. In either case the process may be retarded by cold weather. The ammonia is of course oxidized to nitrate which is undesirable in other than small amounts in raw water supplies. Removal of ammonia by some other means may therefore prove to be more acceptable. Air stripping of ammonia can give good removals but tends to be very costly because of the large air flows required and the chemicals necessary to produce the high pH. Ion exchange using clinoptilolite to remove ammonia is possible although the exchange capacity of the material is relatively low. Some success has been achieved using ion-exchange resins to remove nitrate although the available resins are not ion selective so that the cost and efficiency of this process depends upon the other ions present in the raw water.

High removals of nitrate produced by biological nitrification can be achieved by mixing recycled effluent with incoming settled sewage in a low DO, or anoxic, environment. In these circumstances oxygen is removed from the nitrate by biological denitrification so that most of the nitrate is converted to nitrogen which escapes to the atmosphere. Full-scale operation of activated-sludge plants with an anoxic zone at the inlet end has shown 80 % or higher nitrate removals with a reduction in power costs due to the absence of aeration in the anoxic zone. Similar biological denitrification to remove nitrates from waters low in organic matter can be achieved in submerged bacteria beds or fluidized beds with the oxygen demand being provided by the addition of a cheap organic material such as methanol.

Phosphates

Phosphates in sewage effluents are fairly easily removed by precipitation with aluminium sulphate or lime and this is best carried out at the sewage works where removal should obviate possible algal problems in the receiving water. Some attempts have been made to operate activated-sludge systems under conditions which promote excess uptake of phosphorus but unless this additional material can be retained in the solids during sludge treatment the process is of little value.

Micro-organisms

Conventional disinfection with chlorine or other agents is normally able to deal satisfactorily with very large concentrations of bacteria in raw waters. The removal of soluble organic matter and ammonia will be beneficial in achieving disinfection without the need for excessive doses of disinfectants. Although it is believed that viruses are fairly readily inactivated by most chemical disinfectants it is known that large numbers of viruses can be found in wastewaters. Since the number of viral particles necessary for infection is often small it is clear that careful monitoring for viral content would be desirable in any re-use or reclamation operation involving sewage effluent.

19.4. Physico-chemical Treatment of Wastewater

In this discussion of tertiary treatment and water reclamation it has been assumed that the sewage had been treated by conventional processes with biological units playing a major role in the removal of organic compounds. These processes are well developed and reliable in most situations so that alternative processes do not seem likely to supplant them to any major extent. Nevertheless a considerable amount of research has recently been undertaken, mainly in the USA, to determine the performance characteristics of physico-chemical treatment plants using chemical coagulation and precipitation followed by filtration and adsorption. Such processes have been claimed to produce effluents of 10 mg/l BOD and 20 mg/l COD from relatively weak US sewages. Work in countries like the UK with lower water consumptions suggests that physico-chemical treatment is not competitive with conventional methods in terms of effluent quality and overall cost for normal wastewaters. The costs of chemicals and energy for physico-chemical processes tend to be high and the large volume of sludge produced could also cause problems.

References

1. PUGH, N. J., Treatment of doubtful waters for public supplies. *Trans. Inst. Wat. Engrs*, **50**, 1945, 80.
2. METZLER, D. F. *et al.*, Emergency re-use of reclaimed water for potable supply at Chanute, Kan. *J. Am. Wat. Wks Assn*, **50**, 1958, 1021.

Further Reading

ARGO, D. C., Cost of water reclamation by advanced wastewater treatment. *J. Wat. Pollut. Control Fedn*, **52**, 1980, 750.

BAILEY, D. A., JONES, K. and MITCHELL, C., The reclamation of water from sewage effluents by reverse osmosis. *Wat. Pollut. Control*, **73**, 1974, 353.

BANKS, N. B. and BUTWELL, A. J., The Coleshill Advanced Waste-Water Treatment Plant. *Wat. Pollut. Control*, **80**, 1981, 559.

BENNEWORTH, N. E. and MORRIS, N. G., Removal of ammonia by air stripping. *Wat. Pollut. Control*, **71**, 1972, 485.

BOUCHER, P. L. *et al.*, Use of ozone in the reclamation of water from sewage effluent. *Inst. Pub. Hlth Engrs J.* **67**, 1968, 75.

BURLEY, M. J. and MELBOURNE, J. D., Desalination. In *Developments in Water Treatment*, 2 (Ed. Lewis, W. M.), Applied Science Publishers Ltd., Barking, 1980.

COOPER, P., Physical and chemical methods of sewage treatment. Review of present state of technology. *Wat. Pollut. Control*, **74**, 1975, 303.

COX, G. C. and HUMPHRIS, T. H., The use and re-use of sewage effluent. *Wat. Pollut. Control*, **75**, 1976, 413.

CROOK, J., Health aspects of water reuse in California. *J. Envr. Engng Div. Am. Soc. Civ. Engrs*, **104**, 1978, 601.

DEAN, R. B. and LUND, E., *Water Reuse*, Academic Press (London) Inc., London, 1981.

GARRISON, W. E. and MICLE, R. P., Current trends in water reclamation technology. *J. Am. Wat. Wks Assn*, **69**, 1977, 364.

GAUNTLETT, R. B., Removal of nitrogen compounds. In *Developments in Water Treatment*, 2 (Ed. Lewis, W. M.), Applied Science Publishers Ltd., Barking, 1980.

HIRST, G. and ROCK, B. M., The Pudsey Project: An experiment in direct re-use of sewage effluent for wool textile processing. In *Sewage Effluent as a Water Resource*, IPHE, 1973, p. 81.

HOBBS, J. M. S., Sea water distillation in Jersey and its use to augment conventional water resources. *J. Instn Wat. Engrs Scits*, **34**, 1980, 115.

McCARTY, P. L., Organics in water—an engineering challenge. *J. Envr. Engng Div. Am. Soc. Civ. Engrs*, **106**, 1980, 1.

MERCER, B. W. *et al.*, Ammonia removal from secondary effluents by selective ion exchange. *J. Wat. Pollut. Control Fedn*, **42**, 1970, R95.

MILLER, D. G. and SHORT, C. S., Treatability of River Trent water. In *Symposium on Advanced Techniques in River Basin Management: The Trent Model Research Programme*, IWE, 1973, p. 43.

NAUGHTON, J., Water Research Centre unravels the nitrate knot. *Surv.* **144**, 1974, No. 4285, 12.

ROOK, J. J., Production of potable water from a highly polluted river. *Wat. Treat. Exam.* **21**, 1972, 259.

SHORT, C. S., Removal of organic compounds. In *Developments in Water Treatment*, **2** (Ed. Lewis, W. M.), Applied Science Publishers Ltd., Barking, 1980.

SOUTH TAHOE PUBLIC UTILITY DISTRICT, *Advanced Wastewater Treatment as Practiced at South Tahoe*, Environmental Protection Agency, Water Quality Office, Washington D.C., 1971.

SPARHAM, V. R., Improved settling tank efficiency by upward flow clarification. *J. Wat. Pollut. Control Fedn*, **42**, 1970, 801.

TAYLOR, E. W., Improvement in quality of sewage works effluent after passing through a series of lagoons. *Inst. Pub. Hlth Engrs J.* **65**, 1966, 86.

TEBBUTT, T. H. Y., An investigation into tertiary treatment by rapid filtration. *Wat. Res.* **5**, 1971, 81.

VANDYKE, K. G., Microstraining in water pollution control. *Effl. Wat. Treat. J.* **11**, 1971, 373.

VAN VUUREN, L. R. J., CLAYTON, R. J. and VAN DER POST, D. C., Current status of water reclamation at Windhoek. *J. Wat. Pollut. Control Fedn*, **52**, 1980, 661.

WAGGOTT, W. and BAYLEY, R. W., The use of activated carbon for improving the quality of sewage effluent. *Wat. Pollut. Control*, **71**, 1972, 417.

WALLBANK, T. E. and PARRY, T. A., Pilot-scale advanced waste-water treatment at Davyhulme. *Wat. Pollut. Control*, **81**, 1982, 127.

WINFIELD, B. A., The performance and fouling of reverse-osmosis membranes operating on tertiary-treated sewage effluent. *Wat. Pollut. Control*, **77**, 1978, 457.

CHAPTER 20

Water Supply and Sanitation in Developing Countries

IN DEVELOPED countries the public expect, and usually get, a high-standard water-supply service and efficient collection, treatment and disposal of wastewaters. The techniques for pollution control are, in general, well developed and since populations are in low or zero-growth states demands on water resources are not usually excessive. The picture is very different in developing countries where some 2000 million people are without safe water and adequate sanitation; this means that about 70% of the population in these parts of the world lack basic facilities. The cost, in both monetary and manpower terms, of rectifying the situation will be large. Although for urban areas throughout the world the ultimate aim may be to provide a developed-country level of service the provision of services at this level for the millions in rural areas is unrealistic. Thus whilst the principles and processes discussed in this book are suitable for application in developed countries and as a target for all urban areas, more appropriate techniques must be adopted for rural communities in low-income areas. It is important to appreciate that to reduce the toll of water-related disease improvements must be made in both water supply and sanitation although unfortunately sanitation is often neglected in favour of the more attractive water-supply activities. It is equally important to understand that the construction of sophisticated developed-country-style water and waste-water treatment facilities, often favoured by donor governments and agencies, is of little value if the appropriate operation and maintenance back-up is not also provided.

213

20.1. Sources of Water

In the developed countries it is normal to provide at least some degree of treatment for water from any source whereas for rural schemes in developing countries treatment will not be feasible in many circumstances. It is thus necessary to consider water sources in relation to what is likely to be the most important quality parameter, that of bacteriological quality.

Rainwater

With reasonably reliable rainfall the collection and storage of runoff from roofs can give a quite satisfactory source of water provided that the first flush of water from a storm, which is likely to be contaminated by bird droppings, etc., can be diverted away from the storage tank. With irregular rainfall, the size and cost of storage tanks may be large and unless the tanks are protected from contamination and the entry of mosquitoes health problems can arise. Depending upon the intensity of the rainfall and the efficiency of the gutter and down pipe system, between 50 and 80 % of the rainfall may be collected.

Springs

Spring water is normally of good quality provided that it is derived from an aquifer and is not simply the discharge of a stream which has gone underground for a short distance. It is important to maintain this good quality by protecting the spring and its surroundings from contamination by man and animals. A collecting tank should be constructed to cover the eye of the spring and prevent debris being washed into the supply.

Tube Wells

Because of the natural purification, which removes suspended matter such as bacteria, groundwaters are usually of good bacteriological quality. Care must, however, be taken to ensure that sanitation practices, or the lack of them, do not cause groundwater contamination. Driven wells formed by well points in suitable ground conditions are relatively cheap although they often have a limited life due to corrosion of the tube and clogging of the perforations with soil particles. In sandy soils, tube wells made from plastic pipes may be rapidly driven by jetting. Bored wells can be produced by hand auger or by machine. Small diameter wells (40–100 mm) are normally fitted with simple hand pumps at the surface when the water table is sufficiently close to the surface and a great deal of work is in progress to produce a sturdy reliable hand-pump design. For deeper water table sites, where a surface pump will have insufficient lift, the pump must be placed down the well which usually requires a larger diameter bore and thus increases the cost. The head of a tube well should be provided with a suitable cap to prevent the entry of contaminated surface water.

Hand-dug Wells

In many parts of the world hand-dug wells, 1–3 m diameter, are the traditional sources of water in rural areas. Depending upon the depth of the water table these wells may be as much as 30 m deep and pose considerable hazards during construction since the risk of collapse is often high. This hazard can be greatly reduced by the use of precast concrete rings which sink as excavation proceeds and provide a permanent lining. In old wells the water quality often leaves much to be desired because of contamination due to the entry of surfacewater, spillage from containers and the deposition of debris. It is important that the site of the well is such as to avoid the entry of potentially contaminated groundwater and that a watertight lining extends for 3–6 m below the surface. The well head should have a headwall and drainage apron so that any surfacewater and/or spillage cannot gain entry to the well. These features are particularly important in areas where guinea-worm infections are endemic. Where possible, a pump should be used for water abstraction thus permitting the use of a fixed cover on the well which will further reduce the risk of contamination of the water in the well.

Infiltration Galleries

A porous collector system using open-jointed pipes in a gravel and sand filled excavation can be used to intercept high groundwater tables and give a further degree of filtration to the water. A similar arrangement can be usefully employed when abstracting water from rivers and lakes.

Surfacewater Abstraction

The traditional developed-country sources of water in the form of rivers and lakes exist in many parts of the world but in tropical countries the quality of surfacewaters is often poor so that for rural supplies it is advisable to use surfacewater only as a last resort.

The basic characteristics of suitable rural water sources are shown in Fig. 20.1. The undoubted attractions of groundwater supplies as regards bacteriological quality have resulted in many rural water schemes based on tube wells and it is vital to appreciate that it is only possible to abstract groundwater at a rate not exceeding the natural recharge. Disregard of this basic principle has led to falling groundwater tables and exhaustion of wells in a number of developing countries.

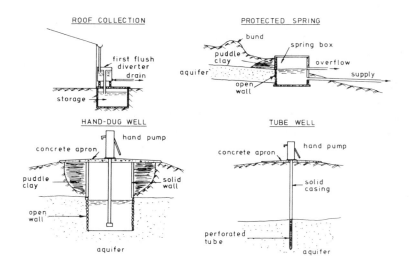

FIG. 20.1. Rural water sources.

20.2. Water Treatment

In the previous section emphasis was placed on the need to obtain water from sources likely to be free from harmful levels of contamination. This is particularly important for rural water supplies since treatment processes greatly increase the cost of providing water and unless operation and maintenance skills are available the treatment is soon likely to fail. There is no such thing as a maintenance-free treatment process and for developing-country situations, water treatment should not be adopted unless its use is unavoidable. In this context it is not realistic with small rural supplies to adopt the normal requirement of no *E. coli* in 100 ml of water since this would imply the need for treatment from many sources. There is considerable evidence that waters with up to 1000 *E. coli* per 100 ml can be supplied to rural communities with little or no health hazard. Indeed, the provision of such a poor-quality (by developed-country standards) water could well bring considerable improvements in health in the many areas where water-washed diseases are the major source of ill health. When considering alternative sources of water it is worth remembering that a good-quality source some distance away from the community could possibly be conveyed to the community at a lower long-term cost than that for treating a nearer but poorer-quality source and the former solution is certainly likely to be more reliable.

If there is no alternative to the provision of treatment every effort must be made to keep the treatment as simple as possible to try to ensure low cost, ease of construction, reliability in operation and to enable operation and maintenance to be satisfactorily undertaken by local labour. Failure to satisfy these basic aims will almost certainly produce many problems and will often lead to the abandonment of the scheme with reversion to the traditional unimproved sources.

Storage

Storage can provide a useful measure of purification for most surface-waters although it cannot be relied upon to produce much removal of turbidity. The disinfecting action of sunlight normally gives fairly rapid reductions in the numbers of faecal bacteria. To obtain the maximum benefit from storage it is important to ensure that short-circuiting in the basin is prevented by suitable baffles. A disadvantage of storage in hot climates is the considerable evaporation losses which can occur. The design of storage facilities should be such as to prevent the formation of shallow areas at the edges which could provide mosquito-breeding sites. If the settlement provided by storage does not give sufficient removal of suspended matter the choice for further treatment involves consideration of chemical coagulation and/or filtration techniques.

Coagulation

The use of chemicals for coagulation brings a further level of complexity to the treatment process and should only be adopted if the necessary skills are available locally. Chemical coagulation will only be successful if the appropriate dose can be determined and then applied to the water in such a manner as to ensure adequate mixing and flocculation. The most useful form of chemical feeder is one based on hydraulic control of a solution such as the Marriotte vessel which provides a constant rate of discharge regardless of the level in the storage container. The coagulant must be added at a point of turbulence such as a weir or in a baffled channel and flocculation is best achieved in a baffled basin connected to a settling basin. In practice it is difficult to prevent floc carry-over from the settling basin so that the output water quality may at times have fairly high turbidity levels. A more satisfactory water may therefore arise by omitting the coagulation stage and proceeding directly with filtration.

Filtration

Although some simplified types of rapid sand filter are available the slow sand filter is likely to be the most satisfactory form of treatment process for many developing-country installations. Slow filtration is able to provide high removals of many physical, chemical and bacteriological contaminants from water with the advantages of simplicity in construction and use. No chemicals are required and no sludge is produced. Cleaning by removal of the top surface at intervals of a month or more is labour intensive but this is not normally a problem in developing countries. The relatively large area requirements for a slow filter are unlikely to cause problems for small supplies and the cost of construction can be reduced by using locally available sand or substitute materials such as rice husks. It is usually possible to operate slow filters at rates of about $5\,m^3/m^2\,d$ with raw water turbidities of up to 50 NTU whilst producing a filtrate of < 1 NTU.

Disinfection

If disinfection is required the previously stated problems related to chemical dosing must again be faced. Chlorine is the only practicable disinfectant but the availability of the gaseous form is likely to be restricted and in any event the hazards of handling chlorine gas make it unsuitable for rural supplies. A more suitable source of chlorine for such installations is bleaching powder which is about 30 % available chlorine and is easy to handle although it loses its strength when exposed to the atmosphere and to light. High-test hypochlorite (HTH) in granular or tablet form has a higher available chlorine content (70 %) and is stable in storage but more costly. Sodium hypochlorite solution is another possible source of chlorine. Dosage should be by means of the same type of simple hydraulic feeder as discussed for coagulants. For very small supplies simple pot chlorinators using a pot or jar with a few tiny holes and filled with a mixture of bleaching powder and sand can give a chlorine residual for about 2 weeks before recharging is necessary. The attainment of the correct dose of chlorine is important since too low a dose will give a false sense of security and too high a dose will give a chlorine taste to the water which will probably result in rejection of the supply by the consumers in favour of traditional sources.

20.3. Sanitation

The relationships between a considerable number of water-related diseases and the presence in the environment of excreta from people suffering from these diseases are well established. It could indeed be argued that the sanitary disposal of human excreta is more important in a health context than the provision of a safe water supply. Even in the presence of good-quality water, direct faecal−oral contact can maintain high levels of incidence of diseases such as typhoid and cholera. It would therefore seem important to make every effort to prevent faecal contamination of water sources as a primary objective since the treatment of an already polluted water can be costly and, particularly on a small scale, is unlikely to have high reliability.

Excretion is inevitably a highly personal process and as such is largely governed by the sociological patterns in a particular community. A vital first step in any sanitation programme is therefore to gain a full understanding of current excretion practices and of the likely acceptability of possible alternatives. It is generally true in rural areas that excreta disposal is far more complex socially than it is technically. A purely engineering solution may well be quite unsatisfactory because its sociological implications have not been examined. Improvements in public health do not necessarily follow the installation of a sanitation system since unless the new facilities are correctly used and given the appropriate level of maintenance little benefit will arise.

When considering the various types of sanitation systems a basic differentiation can be made between dry systems which essentially handle only faeces possibly with some urine, and wet systems which handle faeces, urine and sullage (the liquid wastes from cooking, washing and other household operations). A simple classification of sanitation methods, the more important of which are shown schematically in Fig. 20.2, would be:

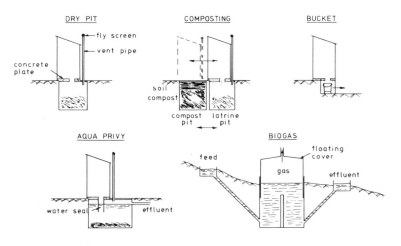

FIG. 20.2. Simple sanitation systems.

Dry, on-site treatment and disposal
  trench and pit latrines, composting latrines.
Dry, off-site treatment and disposal
  bucket or vault latrine with collection service and central treatment
  facility.
Wet, on-site treatment and disposal
  wet pit, aqua privy, septic tank, biogas, land disposal.
Wet, off-site treatment and disposal
  conventional or modified sewerage and central treatment facility.

Pit Latrines

These provide the simplest form of latrine and are widely used because of
their simplicity, low cost and ease of construction in suitable ground
conditions. The usual form has a pit about 1 m square and 3–4 m deep. A
volume of 0.06–0.1 m³/person year is often used to estimate the life of a pit
and when it is about two-thirds full it is filled in with soil and the
superstructure transferred to a new pit.

If a single pit is to be emptied for further use off-setting it by using an
inclined chute enables more convenient operation. The provision of a vent
pipe from the pit to a height above the top of the superstructure will do
much to reduce smells which otherwise may discourage use of the latrine. A
variant of the pit latrine uses a bored hole 200–400 mm in diameter and
perhaps 6 m deep. The life of such a unit is likely to be less than that of a pit
and fouling of the sides of the bore often produces odour problems.
Unlined pits for excreta disposal must only be used in situations where
there is no danger of groundwater contamination. They should always be
placed downhill of any water source and not within 30 m of a well used for
water supply.

Composting Latrines

In some developing countries the fertilizer value of human excreta may
be a significant factor in crop production so that sanitation methods which
permit excreta re-use for such purposes are appropriate. It is highly
desirable that pathogenic organisms in the excreta are destroyed before the
material is applied to land and crops since otherwise the potential for the

spread of disease is considerable. The required destruction of pathogens can be achieved by composting excreta, vegetable scraps, grass cuttings etc., usually in a batch process with relatively long retention time or possibly in a continuous composting unit with a retention time of a few months. For effective composting a C : N ratio of between 20 and 30 to 1 is necessary and the moisture content must be within the range 40–60 %. Batch composters are typified by the double-vault units widely used in S.E. Asia where the latrine is built on top of two bins which serve in turn as receptacles for faeces, cleaning paper and wood ash. Ash amounting to about one-third of the weight of faeces is normally sufficient to prevent odours. Urine is collected separately and used directly on the land. When a bin is about two-thirds full the contents are levelled, covered with earth and the bin sealed. Defecation is transferred to the second bin and the first bin left for a period of up to 12 months before being emptied. A bin capacity of $0.4\,m^3$ per person is often recommended based on a 1-year cycle of operation.

Although a number of continuous composting systems have been produced in the developed countries their performance in developing countries has not been very satisfactory. The proper operation of a composting toilet needs careful attention, particularly in regard to the moisture content, and they should not be used in preference to a pit latrine unless the requisite degree of supervision can be provided.

Bucket and Vault Latrines

The removal of excreta from latrines in a variety of containers is one of the oldest forms of sanitation and is still widely used in many parts of the world. The traditional bucket latrine utilizes a squatting plate or pedestal set above a metal bucket in a chamber with a door to the outside of the house. The bucket is emptied, usually at night (hence the term nightsoil), into a larger container carried by hand or on a cart to a collection depot or disposal area. The system as normally operated has little to commend it from a hygienic aspect. Spillages during handling occur frequently and the buckets are rarely cleaned so that fly nuisances are common. With improvements such as lids for the buckets during handling, washing and disinfection of the buckets and well-designed and maintained latrines the system could be made more acceptable.

A development of the bucket system uses water-tight vaults below the latrine which are emptied at intervals of about 2 weeks using a suction tanker which may be hand operated or mechanized. With proper design and maintenance such a system can be quite satisfactory but the cost and complexity of the removal system makes its suitability for developing countries rather limited although it is widely used in Japan.

Ultimate disposal of the collected waste is usually achieved by burial in shallow trenches, often hand-dug to a size of $4 \times 1 \times 0.5$ m. These are filled with nightsoil to a depth of about 0.3 m and backfilled with soil. The area can be reused after a period of at least 12 months. Unfortunately in some areas the nightsoil is simply dumped on waste ground with no attempt at burial thus producing highly undesirable environmental conditions.

Wet Pit Latrines

In a number of developing countries the pour-flush water-seal latrine is popular. Water usage to maintain the seal is $1-3$ l per occasion so that the contents of the pit become semi-liquid. Anaerobic digestion of the contents occurs thus reducing their volume to some extent so that a design volume of $0.04-0.06$ m^3/person year should be suitable. The pit is usually placed a short distance away from the latrine and connected to it by a short length of steeply sloping 100-mm pipe. The pit is usually lined with open jointed brickwork to prevent collapse but to allow percolation of the liquid contents into the surrounding ground. The water seal means that fly and odour nuisances are prevented so that such a latrine is suitable for indoor installation. Success of a wet pit system depends upon a relatively low level of water usage and the presence of suitable ground conditions to allow escape of the liquid without causing groundwater pollution.

Aqua Privies

An aqua privy consists of a small tank situated below a latrine and discharging an anaerobically treated liquid effluent which must be suitably disposed of to complete the system. The latrine plate or pan must have a submerged down pipe or water trap to maintain anaerobic conditions in the tank, prevent the escape of odours and stop the entry and egress of insects. A typical tank capacity would be $0.12 \, \text{m}^3$/person with a sludge accumulation of about $0.04 \, \text{m}^3$/person year. The volume of effluent discharged is likely to be about $6 \, \text{l}$/person day. A major operating problem which has been found with many aqua-privy installations is that insufficient water is added to maintain the water seal so that the latrine becomes unattractive to the user. Unless evaporation and leakage losses can be made up the liquid level in the tank falls, breaking the seal and giving odour and insect problems. In some installations sullage water is piped to the tank to provide the necessary make-up volume, in this case an additional $0.5 \, \text{m}^3$ capacity should be provided to allow for the sullage.

Septic Tanks

As indicated above, aqua privies are basically simplified septic tanks which are widely used throughout the world in many rural areas of developed countries. As anaerobic units they provide the removal of much of the suspended matter from sewage, the deposited solids then digesting with consequent release of some of the organics in soluble form. A typical septic tank will remove about 45% of the applied BOD and about 80% of the incoming SS. The effluent will contain large numbers of bacteria and concentrations of 10^6 E. coli/100 ml are not uncommon in the discharge from a septic tank. Solids accumulations of around $0.05 \, \text{m}^3$/person year are often assumed and the sludge must be removed at intervals of $1-2$ years. The sludge is very strong with BOD and SS levels of between $10\,000$ and $50\,000$ mg/l. Typical design criteria are a minimum liquid retention time of 3 days and for populations of between 4 and 300, UK practice is to provide a volume in m^3 of $(0.18 \times \text{population} + 2)$. Again it is important to provide suitable inlet and outlet arrangements to ensure a water seal and prevent the discharge of the scum layer which usually exists on the top of the liquid.

Biogas

The utilization of methane produced by anaerobic digestion of wastewater sludges is well established in developed countries, normally accompanied by a great deal of complex and expensive plant. However, any container holding putrescible organic matter will develop anaerobic conditions with the consequent production of methane. Thus an individual dwelling aqua privy or septic tank will develop a small amount of methane but the volume is insufficient for any significant use. In rural areas in developing countries the collection of solid wastes from man and animals in a simple container can produce sufficient gas for domestic lighting and cooking requirements and the digested sludge has a high nitrogen content making it a useful fertilizer. Many biogas installations constructed from simple materials are in use in the Far East. Most of these plants are quite small, 1–5 m³ capacity, and are operated by individual farmers. The daily gas requirement for cooking is about 0.2 l/person day and this requirement can probably be satisfied by digestion of human excreta plus the excreta from a single cow or similar beast. Design loadings for biogas units are around 2.5 kg VS/m³ d with a nominal retention time of 20 days or more.

On-site Effluent Disposal

For all wet sanitation systems safe disposal of the liquid effluent is an inherent part of the system. The effluent from such systems is likely to be relatively low in SS but will probably have a high organic content and large numbers of bacteria so that indiscriminate release to the environment

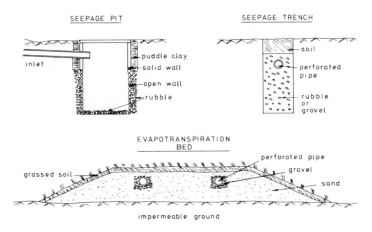

FIG. 20.3. On-site effluent-disposal systems.

would create health hazards. In many circumstances the most satisfactory method of disposal is into the ground via seepage pits, drainfields or evapotranspiration beds (Fig. 20.3). The applicability of the various methods is related to soil permeability, groundwater levels, and proximity to buildings. If sub-surface disposal is not possible, further treatment of the effluent in a facultative oxidation pond or possibly a simple bacteria bed may be feasible for a community. Septic tanks, aqua privies and soakaways should normally be sited at least 30 m away from wells and boreholes, on the downstream side of the flow and similar distances should be used in relation to sites near surface water sources.

A soakaway may be in the form of a seepage pit, with porous construction, in sites where the soil is highly permeable. The pit should be of similar size to the aqua privy or septic tank which it serves. In most soils, drainfields provide the most satisfactory form of sub-surface disposal. The system comprises seepage trenches containing open-jointed or perforated pipes surrounded by stone fill and topped by an earth backfill. Most of the percolation occurs through the sides of the trench and a reasonable loading for many soils is $10 l/m^2 d$, based on sidewall area.

In areas where the ground is impermeable an artificial soakage area or evapotranspiration bed can be provided in a suitable depression or above the general ground level in the form of a mound. The bed or mound is made of coarse sand and gravel surrounding the pipes and topped with a layer of soil supporting a fast-growing grass. Liquid losses from such an area will probably be about 80 % of the evaporation from a free water surface in the same locality.

Whilst sub-surface disposal methods can be quite satisfactory in low population-density areas the possible hazard to groundwater quality must be recognized and in urban areas it is unlikely that sufficient land area will be available for the methods to operate correctly.

Off-site Treatment and Disposal

In urban areas it may be necessary to install sewerage systems to collect all liquid wastes and convey them to a treatment facility. Unless a reliable water supply is available a conventional sewerage system will have many problems due to solids deposition in low flows, hydrogen sulphide production and consequent corrosion effects being particularly troublesome. The capital cost of a conventional sewerage system is very high and the disruption caused by its construction in a congested urban area must also be considered. In some areas a modified sewerage system has been adopted to collect the effluent from individual septic tanks and aqua privies. In this situation the flows are likely to be relatively low and since the bulk of the solids have been removed, small diameter pipes laid at shallow gradients will suffice, greatly reducing construction problems and costs. It must be appreciated that such a modified sewerage system will only work satisfactorily if the individual tanks are regularly desludged.

With a sewerage system it will be necessary to provide some form of central treatment facility to ensure that the effluent can be discharged without causing significant environmental damage. In most developing-country situations the best form of wastewater treatment will probably be provided by facultative oxidation ponds which are simple to construct and operate. They do, however, require considerable amounts of land so that in urban areas this may be a problem. Care must be taken to prevent shallow water and vegetation at the edges providing conditions attractive for mosquitoes and other insects. The large algal growth in such ponds usually means that the SS in the effluent are relatively high due to escaping algae. In some areas the possibility of protein recovery from algal matter may be worth consideration. If conditions are unsuitable for oxidation ponds, bacteria beds or activated-sludge units may be necessary but these relatively complex installations should only be adopted if the appropriate level of operational and maintenance skills can be assured and that the necessary financial support is also available.

Further Reading

CAIRNCROSS, S. and FEACHEM, R., *Small Water Supplies*, Ross Institute, London, 1978.

DIAMANT, B. Z., The role of environmental engineering in the preventive control of water-borne diseases in developing countries. *Roy. Soc. Hlth J.* **99**, 1979, 120.

FEACHEM, R., McGARRY, M. and MARA, D. D., *Water, Wastes and Health in Hot Climates*, Wiley, Chichester, 1977.

FEACHEM, R. and CAIRNCROSS, S., *Small Excreta Disposal Systems*, Ross Institute, London, 1978.

IWUGO, K. O., Basic data requirements for the planning and implementation of appropriate sanitation technology in Africa. *Pub. Hlth Engr*, **18**, 1980, 26.

KALBERMATTEN, J. M., Appropriate technology for water supply and sanitation: Build for today, plan for tomorrow. *Pub. Hlth Engr*, **9**, 1981, 69.

KLEIN, R. L., CANNON, D. E. and PRYNN, P. J., Water and sewerage appraisals in developing countries and their pertinence to UK practice. *Pub. Hlth Engr*, **4**, 1976, 76.

MARA, D. D., *Sewage Treatment in Hot Climates*, Wiley, Chichester, 1976.

MANN, H. T. and WILLIAMSON, D., *Water Treatment and Sanitation*, Intermediate Technology Development Group, London, 1978.

PACEY, A. (ed.), *Sanitation in Developing Countries*, Wiley, Chichester, 1978.

PICKFORD, J., Sewerage for developing countries. *Effl. Wat. Treat. J.* **18**, 1978, 119.

PICKFORD, J., Control of pollution and disease in developing countries. *Wat. Pollut. Control*, **78**, 1979, 239.

WHITE, G. F., BRADLEY, D. J. and WHITE, A. U., *Drawers of Water*, University of Chicago Press, Chicago, 1972.

WOLMAN, A., Sanitation in developing countries. *Pub. Hlth Engr*, **6**, 1978, 32.

Various Authors, *Appropriate Technology for Water Supply and Sanitation*, **1–12**, World Bank, Washington, 1980–82.

Index

229